Binary Sequences

G. Hoffmann de Visme

Head of Electronic and Electrical Engineering,
North Staffordshire Polytechnic; formerly
Lecturer at University of Manchester
Institute of Science and Technology

The English Universities Press Ltd

Editor's Foreword

Many engineers who today find themselves working with binary sequences will not have had the benefit of hearing discussion about them in the lecture theatre. Like many of the newer subjects, if it is to be understood at all it must be understood with the help of books on the subject. The difficulty with such a subject as binary systems is that it can be made to appear mathematically rather formidable. This book will meet the needs of those who would like to have a book about the subject which was reasonably simple in presentation, but not at the cost of completeness. It is hoped that this book fulfils that specification.

J. H. CALDERWOOD

ISBN 0 340 15582 5 (boards edition)
0 340 063491 (Unibook edition)

First printed 1971

The English Universities Press Ltd
St Paul's House, Warwick Lane, London EC4P 4AH

Printed and bound in Great Britain for The English Universities Press Ltd
by Richard Clay (The Chaucer Press) Ltd, Bungay, Suffolk

Preface

The properties of binary sequences, and in particular of the so-called pseudo-random sequences, have been the subject of intensive study for a number of years. This is due firstly to their intrinsic mathematical interest, and secondly to the fact that binary sequences possessing very useful properties can easily be generated by means of simple logical circuits. A very large numbers of papers have been published on sequences and their application, and three books consisting of collections of some of the more important papers have been produced, but as yet no book has appeared in which the main properties and theory of binary sequences are described in a simple and at the same time fairly complete form. The present book is an attempt to fill this gap. It is intended to enable engineers who are not familiar with Galois Field Theory and other mathematical sophistications to make sense of the published literature, which is often couched in somewhat abstruse language. Indeed no mathematical knowledge beyond A-level standard is assumed. With this purpose in mind the discussion is everywhere reinforced by simple examples, and conventional proofs, which are for the most part no more than generalizations of examples, are only given when appropriate. For the benefit of readers who may not be familiar with logic notation or certain processes in number theory, a list of definitions is included at the start of the book.

No originality is claimed beyond that of the treatment, and no new results are propounded. The author has relied entirely on published work, notably by D. A. Huffman (1955), S. Golomb (1956), B. Elspas (1958), and N. Zierler (1959). The bibliography includes over fifty references, but these form only a fraction of the whole. The reader, however, will find no difficulty in tracing the remainder from the references quoted by the authors of the papers mentioned here.

Unlike many new ideas which burst on the scene and are followed by a flood of papers, only to sink back into oblivion soon after, binary sequences have without doubt come to stay as they have proved their usefulness in many applications, some of which are described in some detail at the end of this book.

G. HOFFMANN DE VISME

Manchester

Contents

Editor's Foreword ii

Preface iii

Definitions vii

Introduction: Binary Systems 1

Chapter One: The Linear Binary Sequence Filter 4
- Synthesis of Linear Filters 6
- Realizability Restrictions on Linear Transfer Functions 9
- Behaviour of Different Realizations of a Given Transfer Function 10

Chapter Two: The Feedback Shift Register 15
- Periodicity 16
- State Transitions—the Complete State Diagram 18
- Condition for Separate Cycles 20

Chapter Three: The Cycling Register 22
- The Mobius and Euler Functions 22
- Number of Cycles Produced by a Cycling Register 24
- Successive States of a Cycling Register 26

Chapter Four: Sequences from Linear Feedback Shift Registers 27
- Periods of the Sequences Defined by a Linear Recurrence Relation 27
- Significance of Quotient 30
- Reverse Sequences 31
- Dual Sequences 31
- Cycle Structures Associated with Factorable Polynomials 32
- Polynomial Consisting of Unrepeated Prime Factors 32
- Polynomial Consisting of a Power of a Prime Polynomial 34
- The General Polynomial 36

Chapter Five: Linear Prime Polynomials 37
Number of Prime Polynomials of Degree *n* where $2^n - 1$ is Prime 38
Number of Prime Polynomials of Degree *n*—the General Case 39
Cyclotomic Polynomials 41

Chapter Six: The Shift-and-Add Property and Remote-term Calculation 43
Remote-term Calculation 44
Non-maximal and Product Sequences 46
Mechanization of Polynomial Division 47
Generation of Sequences of Prescribed Period 48

Chapter Seven: Sequence Sampling 52
Application of Sampling to Linear Sequences 55
Complete Coset Structure 56
Sampling of Product Sequences 57

Chapter Eight: Noise-like Properties of *m*-Sequences 58
Criteria of Randomness 58
Randomness Properties of *m*-sequences 62
Cross-correlation 64
Fourier Analysis of *m*-sequences 66
Waveforms Derivable from Maximal Shift Registers 68

Chapter Nine: Other Forms of Binary Sequences 72
Two-level Sequences 72
Levels Giving Zero Off-peak Autocorrelation 74
Number of Runs in a Period 75
Relation Between Two-level Sequences, Hadamard Difference Sets, and Hadamard Designs 75
Known Forms of Two-level Sequences 78
A Distance Property of Pseudo-random Sequences 80
Barker Sequences 81
Inverse Repeat Sequences 82
Periodic Sequence Generation Using Shift Registers 83
The Full-length Sequence 85
Non-linear Feedback Shift Register Sequences 89
Representation of Feedback Shift Registers 91

Chapter Ten: Applications of Binary Sequences 95
Binary Codes 95
Radar 96
Communication 99
Evaluation of System Parameters 102
Theory of Correlation Technique 103
Periodic Input Disturbance 106
System Noise 107
System Drift 108
Conclusion 109

Appendix: Moments of Binary Sequences having Multipliers 110

Bibliography 113

Index 117

Definitions

The following is a list of functions, symbols, and theorems used in this book. Those not included here are defined as necessary in the text.

Logic Functions and Symbols

The following are functions and symbols applicable to the binary alphabet, i.e. to systems and expressions in which each variable or 'literal' is either 1 or 0.

AND function

If a, b, c, etc. are binary variables, a AND b AND c AND . . . = abc . . . = 1 if all the variables are 1, and = 0 if any or all of the variables are 0. The symbol for the AND gate is:

OR function

If a, b, c, etc. are binary variables, a OR b OR c OR . . . = $a + b + c + \ldots$ = 0 if all the variables are 0, and = 1 if any or all of the variables are 1. The symbol for the OR gate is:

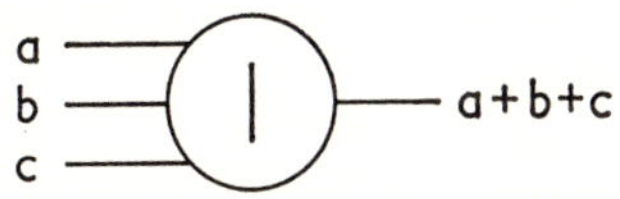

NOT function

If a is a binary variable, NOT $a = \bar{a}$ (sometimes written a') is 1 if a is 0 and 0 if a is 1. The symbol for the NOT gate is:

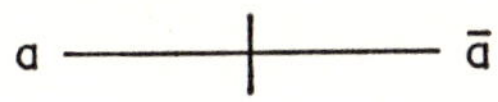

NAND function

The NAND function is the AND function followed by the NOT function. The symbol for the NAND gate is:

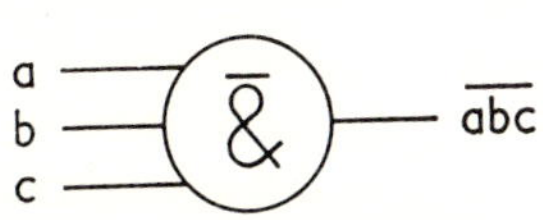

NOR function

The NOR function is the OR function followed by the NOT function. The symbol for the NOR gate is:

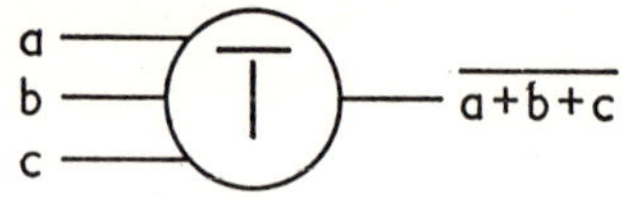

If only one input is used, the NAND gate and the NOR gate function as NOT gates.

Modulo-two addition

The modulo-two sum $a \oplus b \oplus c \oplus \ldots$ of a number of binary variables is 1 if the number of variables which are 1 is odd and 0 if the number of variables which are 1 is even or zero. Since $a \oplus b \oplus c = (a \oplus b) \oplus c$, modulo-two addition of a number of variables can be effected using two-input modulo-two adders.

For two variables a and b, $a \oplus b$ is 0 if $a = b$, and 1 if $a = \bar{b}$. Also $a \oplus 1 = \bar{a}$. This expression, with the aid of de Morgan's theorem (see below), enables any boolean function of binary variables to be written as an expression containing only AND and modulo-two connectives, and logical 1. The symbol for the modulo-two adder is:

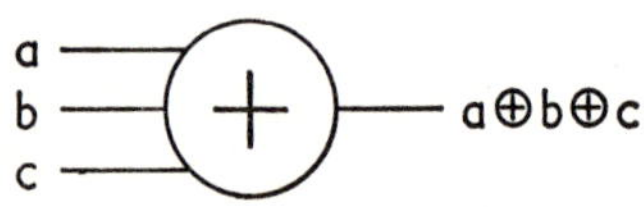

Binary store or memory element

In the context of this book a binary store is a clocked D-type flip-flop represented by the symbol:

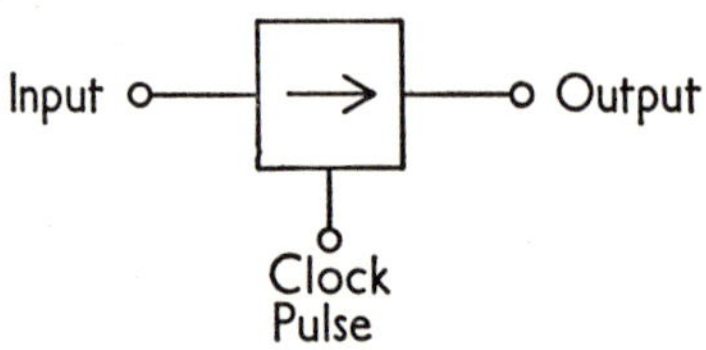

The arrival of a clock pulse causes the output to assume the value of the input at the instant of the clock pulse. The binary store therefore memorises the value (0 or 1) of the input until the next clock pulse arrives. Thus if a_r and b_r are the input and output, respectively, after r clock pulses, $b_r = a_{r-1}$ (frequently written Da_r, where D is the delay operator).

Detached exponent notation

For brevity many expressions and equations are written in detached exponent form. Thus if a_r is a binary variable after r clock pulses, a_r (i.e. D^0a_r) is written as 0, Da_r (i.e. D^1a_r) is written as 1, D^2a_r as 2, etc.

Mathematical Theorems, Functions, and Symbols

The following are mathematical theorems, functions, and symbols used in this book:

de Morgan's theorem

This theorem relates the OR and the AND functions as follows:

$$\overline{a + b + c + \ldots} = \bar{a}\ \bar{b}\ \bar{c} \ldots$$

and

$$\overline{a\,b\,c \ldots} = \bar{a} + \bar{b} + \bar{c} + \ldots$$

where a, b, c, etc. are binary variables.

Mobius function

If n is an integer and d is a divisor of n (1 and n are included as divisors),

$\mu(n) = 1$, if $n = 1$,
$= 0$, if n contains any prime factors raised to a power greater than 1,
$= (-1)^k$, if n is the product of k prime factors none of which is raised to a power greater than 1.

Euler function

If n is an integer the Euler function $\phi(n)$ of n equals the number of numbers less than n and prime to it. The number 1 is regarded as prime to all n.

Fermat's theorem

This theorem states that if n is prime it divides $a^n - a$ ($a = 2, 3, \ldots$). Furthermore if a is not a multiple of n, n divides $a^{n-1} - 1$.

The Fermat–Euler theorem, of which Fermat's theorem is a special case, states that if a and n are relatively prime (neither a nor n need be prime), n divides $a^{\phi(n)} - 1$. If there is a divisor d of $\phi(n)$ for which n divides $a^d - 1$, a is not a primitive root of n. Otherwise a is a primitive root of n.

Congruences

An integer a is said to be congruent to an integer b (modulo m) if m divides $a - b$. Such a congruence is written $a \equiv b \pmod{m}$. Congruences have many of the properties of normal equations, thus if $a \equiv b \pmod{m}$ and $c \equiv d \pmod{m}$, $ac \equiv bd \pmod{m}$. Also $ka \equiv kb \pmod{m}$, $a + k \equiv b + k \pmod{m}$, $a^k \equiv b^k \pmod{m}$, etc.

Correlation

The (second-order) cross-correlation function between two functions $x(t)$ and $y(t)$, where t is usually time, is defined as

$$\phi_{xy}(\tau) = \operatorname*{Lt}_{T\to\infty} \frac{1}{2T}\int_{-T}^{+T} x(t)y(t+\tau)dt.$$

For $x(t)$ and $y(t)$ both of period p this becomes

$$\phi_{xy}(\tau) = \int_{a}^{a+np} x(t)y(t+\tau)dt,$$

where a is arbitrary and n is an integer. Autocorrelation is a special case of cross-correlation where $x(t)$ and $y(t)$ are the same.

Wiener-Khintchine theorem

This theorem relates the Fourier transform $X(w)$ of a time function $x(t)$ with the autocorrelation function $\phi_{xx}(\tau)$ of $x(t)$ as follows:

$$| X(w) |^2 = \int_{-\infty}^{+\infty} \phi_{xx}(\tau)e^{-jw\tau}d\tau$$

$$\phi_{xx}(\tau) = \frac{1}{2\pi}\int_{-\infty}^{+\infty} | X(w) |^2 e^{jw\tau}dw.$$

$| X(w) |^2$ is usually referred to as the power spectral density of $x(t)$.

The Dirac delta function

The Dirac delta function $\delta(x)$ of a variable x is defined as follows:

$$\delta(x - x_0) = 0 \text{ for } x \neq x_0, \text{ but}$$

$$\int_{-\infty}^{+\infty} \delta(x - x_0)dx = 1.$$

Thus $\delta(x - x_0)$ may be represented as a spike at $x = x_0$ of infinite height and infinitesimal width such that the area enclosed by the spike is unity. Note first that $\delta(x)$ has dimensions $[x]^{-1}$, and secondly that since any function $f(x)$ continuous at x_0 is constant and equal to $f(x_0)$ over the width of the spike $\delta(x - x_0)$,

$$\begin{aligned} &\int_{-\infty}^{+\infty} \delta(x - x_0)f(x)dx \\ &= f(x_0)\int_{-\infty}^{+\infty} \delta(x - x_0)dx \\ &= f(x_0). \end{aligned}$$

Introduction

Binary Systems

A large body of theory has been built up concerning systems, both linear and non-linear, in which input and output variables are continuous and (in theory at least) unbounded. In general the behaviour of such a system may be defined by a differential equation relating the input and output variables. Systems containing no energy stores are described by equations free of time derivatives. In these cases the output at any instant is determined by the input at the same instant. The binary system, in which a variable at any point in the system can assume only one of two values, denoted for simplicity 0 and 1, is in some respects similar to the continuous system. For example a binary system in which the outputs at any instant are determined by the inputs at the same instant corresponds to a continuous system containing no energy stores. Binary systems of this kind are usually referred to as static logic, and as in the case of continuous systems, they may be either linear or non-linear.

Relations between the inputs and outputs of static logic are expressed by equations containing the logical functions AND, OR, and NOT. Alternatively, since variables can assume only one of two values, output/input relations may be expressed by means of truth tables. In the case of *linear* logic the equations contain only the exclusive OR, or modulo-two function.

That modulo-two addition is a linear function may be shown as follows: let a binary variable z be the modulo-two sum of variables x and y, i.e. $z = x \oplus y$. If x and y consist of components x_1, x_2 and y_1, y_2 such that $x = x_1 \oplus x_2$ and $y = y_1 \oplus y_2$, then

$$z = (x_1 \oplus x_2) \oplus (y_1 \oplus y_2) = (x_1 \oplus y_1) \oplus (x_2 \oplus y_2).$$

Thus the response to the sum of the input components is the sum of the responses to the input components acting independently. The same does not apply in the case of AND, OR, or NOT functions. For instance, if $z = \bar{y}$ and $y = y_1 \oplus y_2$, linearity would imply that $z = \overline{y_1 \oplus y_2}$ was the same as $\overline{y_1} \oplus \overline{y_2}$, which is not the case.

The behaviour of a binary store or 'memory element' is defined as follows: If x_r and y_r are the input and output of a binary store after r clock pulses (i.e. at epoch r), the arrival of a clock pulse increases r to $r + 1$, and so $y_{r+1} = x_r$. Alternatively, if at any time the input is x_r, the output is

x_{r-1}. For $r = 0$, i.e. before the arrival of any clock pulses, the store output is x_{-1}. This represents the initial (arbitrary) state of the store.

The differential operator D associated with continuous systems containing energy stores corresponds to the delay property of a clocked binary store. Thus a delay operator D is frequently used in this context to express a delay of one epoch, so that Dx is the output of a store whose input is x.

A shift register consists of a number of simultaneously clocked binary stores in cascade. If the input to the first store is x_r at some instant, the outputs of successive stores at the same instant are x_{r-1}, x_{r-2}, x_{r-3}, etc., and the initial state of the register would be the set of output values for $r = 0$, namely x_{-1}, x_{-2}, x_{-3}, etc. After one clock pulse, that is for $r = 1$, the state of the register would be x_0, x_{-1}, x_{-2}, etc., where x_0 was the input to the first store before the arrival of the first clock pulse.

Using the D notation the inputs to the successive stores at any instant would be written x, Dx, D^2x, D^3x, etc. corresponding to the notation used for continuous systems. A term such as D^2x represents two successive delay operations on x, and must not be confused with $(Dx) . (Dx)$, which represents the boolean product Dx AND Dx. This is obvious enough when the symbol D is explicitly included, but detached exponent notation is often used in which D^2x would simply be written 2, which is not the same as 1.1.

The most general binary system having a number of inputs and outputs, a number of simultaneously clocked stores, and static logic is shown in Fig. 1.

The input to each store is a logical function of some or all of the inputs to the system and the outputs of the stores. The simple case mentioned

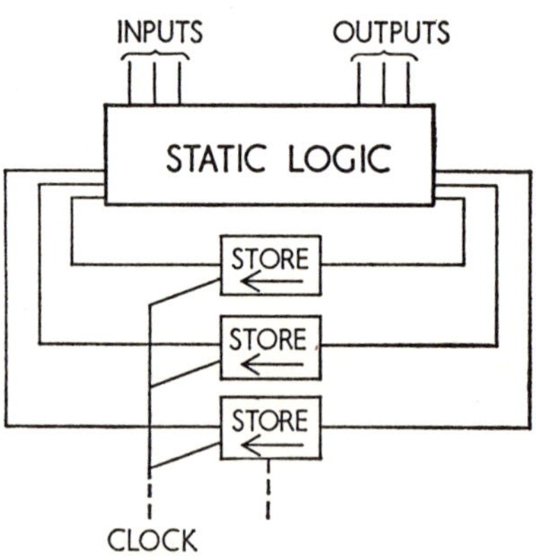

Fig. 1. *General binary system with simultaneously clocked stores*

above of a number of stores in cascade is therefore a special case of this system. Another example is shown in Fig. 2. This is an autonomous, or free-running, system having outputs but no inputs. The clock pulses cause the outputs to run through a normal sequence of three-digit binary numbers. In any conventional counter, of course, counting is achieved far more simply by appropriately interrupting clock pulses, but the example shows that in principle any periodic sequence of multi-bit numbers can be

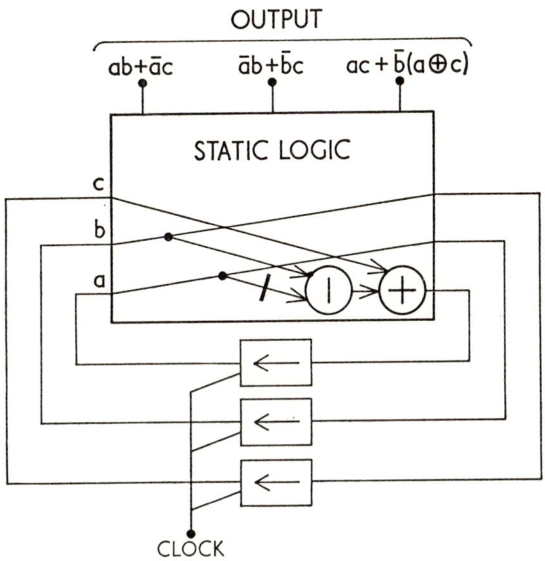

Fig. 2. *Binary counter using simultaneously clocked stores*

derived from a shift register having the same periodicity feeding appropriate static logic.

As has already been indicated, a system is linear if the logic consists only of modulo-two adders, and is non-linear otherwise. Sometimes a subdivision is made comprising systems containing binary stores, linear logic, and invertors (NOT gates) only. In quasi-linear systems of this kind inversion of a variable is equivalent to the modulo-two addition of logical '1' to the variable. The logic may therefore be regarded as linear with the addition of a source of logical '1'.

It is convenient to examine the properties of linear and non-linear systems separately, the former being amenable to analysis in terms of either linear difference equations or characteristic polynomials of the delay operator *D*.

Chapter one

The Linear Binary Sequence Filter

In this chapter linear binary systems having one input and one output are considered, the input being an arbitrary sequence of 1s and 0s, and all stores are simultaneously clocked.

Figs. 1.1 (*a*) and (*b*) show the simplest combinations of store and modulo-two adder in feed-forward and feedback configurations. In the former

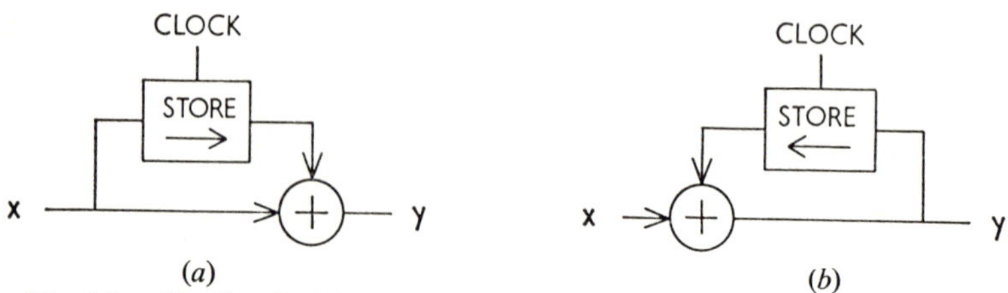

Fig. 1.1. *Simplest feed-forward and feedback circuits*

case $y = x \oplus Dx$, which may be written $(D \oplus 1)x$. In the latter case $y = x \oplus Dy$, i.e. $(D \oplus 1)y = x$, and y may be expressed as $\left(\frac{1}{D \oplus 1}\right)x$. Transfer functions $D \oplus 1$ and $\frac{1}{D \oplus 1}$ may be defined respectively. In terms of input x_r and output y_r at epoch r, difference equations $y_r = x_r \oplus x_{r-1}$ and $y_r \oplus y_{r-1} = x_r$ describe the circuits, respectively. These are first-order equations, and to solve them one initial condition, namely the state of the store, must be prescribed.

A more complicated feed-forward circuit is shown in Fig. 1.2. In difference equation notation, $y_r = x_r \oplus x_{r-1} \oplus x_{r-5}$, and it is seen that the

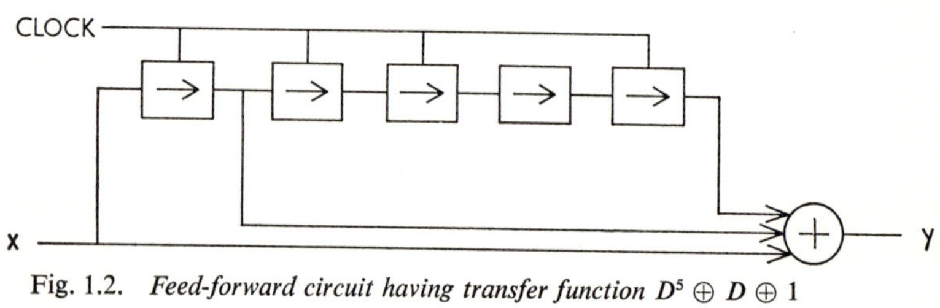

Fig. 1.2. *Feed-forward circuit having transfer function $D^5 \oplus D \oplus 1$*

output at any time is specified either by the input and the state of the stores at that time, or by the input and the five preceding inputs, since these are sufficient to determine the states of the five stores. In D-notation $y = x \oplus Dx \oplus D^5x = (1 \oplus D \oplus D^5)x$, and the transfer function $1 \oplus D \oplus D^5$ may be defined for the filter. In this particular case the same output would

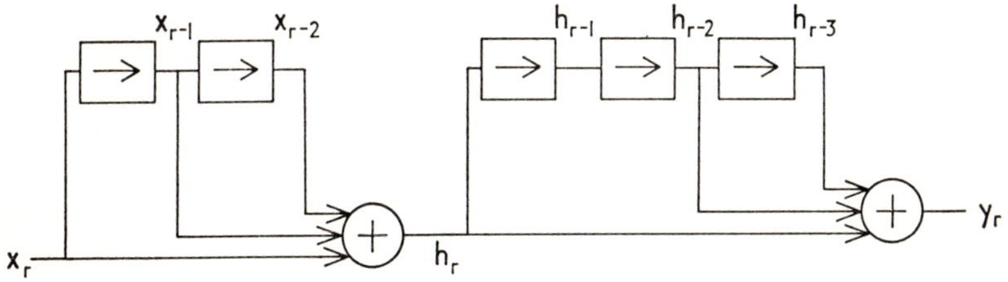

Fig. 1.3. *Alternative realization of transfer function* $(1 \oplus D \oplus D^5)$

result from the circuit of Fig. 1.3 where $h_r = x_r \oplus x_{r-1} \oplus x_{r-2}$. The output y_r is therefore $h_r \oplus h_{r-2} \oplus h_{r-3}$

$$= (x_r \oplus x_{r-1} \oplus x_{r-2}) \oplus (x_{r-2} \oplus x_{r-3} \oplus x_{r-4}) \oplus (x_{r-3} \oplus x_{r-4} \oplus x_{r-5})$$
$$= x_r \oplus x_{r-1} \oplus x_{r-5}, \text{ as required.}$$

The corresponding operation in D-notation is therefore $y = (1 \oplus D \oplus D^2)(1 \oplus D^2 \oplus D^3)x$, and we see that the transfer function in D-notation can be factorized (modulo-two) like a normal algebraic polynomial. In general, linear transfer functions in D-notation can always be treated as algebraic polynomials, but it must be remembered that multiplication of D operators does not correspond to the boolean multiplication of variables, which is a non-linear process.

A feedback circuit corresponding to the above feed-forward circuit is shown in Fig. 1.4. This circuit is defined by the equation $y_r \oplus y_{r-1} \oplus y_{r-5} = x_r$, or $y(1 \oplus D \oplus D^5) = x$, and as before a complete solution

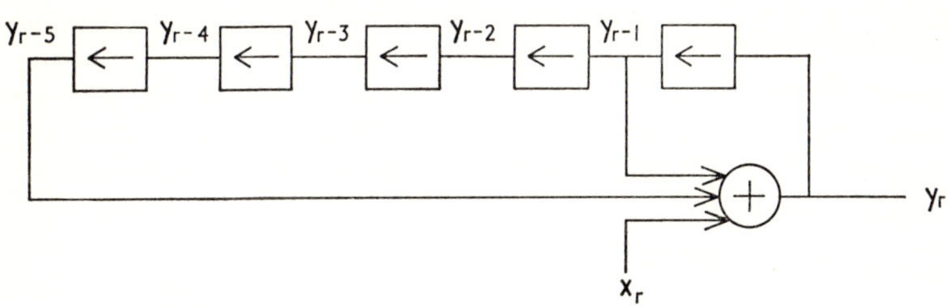

Fig. 1.4. *Feedback circuit having transfer function* $\dfrac{1}{1 \oplus D \oplus D^5}$

requires the states of the five stores at any one time to be defined. In this case the transfer function is $\frac{1}{1 \oplus D \oplus D^5}$ and factorization of the denominator leads to a realization of this transfer function as two feedback circuits in cascade having transfer functions $\frac{1}{1 \oplus D \oplus D^2}$ and $\frac{1}{1 \oplus D^2 \oplus D^3}$. The feed-forward circuit of Fig. 1.2 and the corresponding feedback circuit of Fig. 1.4 are termed inverse circuits, and a cascade connection of the feedback circuit followed by the inverse feed-forward circuit has unity transfer function.

An obvious feature of the above, and indeed any, feedback circuit is that it is autonomous, i.e. it yields a non-trivial output sequence of 1s and 0s, providing the initial states of the stores are not all zero, even when the input is continuously zero. The autonomous behaviour of feedback systems will be discussed in detail later.

Synthesis of Linear Filters

Circuits containing both feed-forward and feedback paths have transfer functions in the form of rational functions of D. As we have seen, these functions may be manipulated algebraically (modulo-two), and various methods exist for synthesizing filters given their transfer function. These methods correspond closely to those for synthesizing continuous-variable linear systems in terms of their transfer function.

A shorthand way of handling linear functions of D is to use detached exponents whereby $1 = D^0$ is written 0, D^1 is written 1, D^2 is written 2, etc., and the modulo-two addition signs are omitted as far as possible. Thus the transfer function

$$\frac{D^8 \oplus D^5 \oplus D^4 \oplus D^2 \oplus D}{D^5 \oplus D^4 \oplus 1} \quad \text{is written simply} \quad \frac{8, 5, 4, 2, 1}{5, 4, 0}.$$

Numerator and denominator of this function have no common factor (modulo-two), and four methods will be described for synthesizing it.

The first method is simply to cascade a feed-forward circuit having the transfer function (8, 5, 4, 2, 1) with a feedback circuit having the transfer function (0/5, 4, 0), where the nought in the numerator is short for D^0, namely the identity operation. The number of stores needed for this synthesis is thirteen.

For the second method the function may be written

$$\frac{(1)\,(7, 4, 3, 1, 0)}{5, 4, 0} = (1)\left(2, 1, 0 \oplus \frac{3, 2}{5, 4, 0}\right),$$

by dividing denominator into numerator. This could be synthesized directly by means of a store feeding into a feed-forward circuit (2, 1, 0) 'in parallel with' circuits (3, 2) and (0/5, 4, 0) in cascade, requiring eleven stores in all. However, the polynomial (5, 4, 0) factorizes into (3, 1, 0) (2, 1, 0), and this enables the function $\frac{3,\,2}{5,\,4,\,0}$ to be split into partial fractions $\frac{0}{3,\,1,\,0}$ and $\frac{0}{2,\,1,\,0}$. This enables the whole transfer function to be synthesized using only eight stores, as shown in Fig. 1.5.

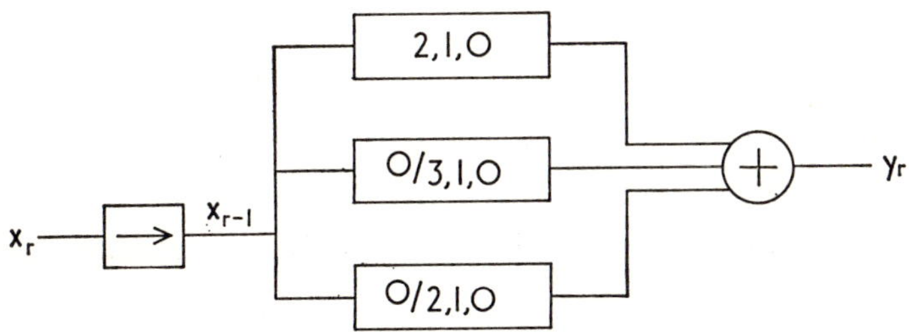

Fig. 1.5. *Partial fraction realization of* $\frac{8,\,5,\,4,\,2,\,1}{5,\,4,\,0}$

The third method is to express the transfer function as a continued fraction. Though uneconomical, this method is instructive and corresponds to the Cauer synthesis of linear networks. Since the transfer function is rational the continued fraction is finite and leads to a finite realization. We have seen (Fig. 1.1 (*b*)) the realization of the function $\frac{1}{1 \oplus D}$. By including feed-forward functions $f_1(D)$ and $f_2(D)$ in the input and feedback loop respectively, the general form $f_1(D)/(1 \oplus f_2(D))$ may be realized. To realize a transfer function by expressing it as a continued fraction, therefore, the function must be expanded into a continued fraction whose successive divisors are all of the form $1 \oplus f(D)$. The above transfer function is expanded into a continued fraction of the required form as follows:

$$\frac{8,\,5,\,4,\,2,\,1}{5,\,4,\,0} = (1)\left(\frac{0,\,1,\,3,\,4,\,7}{0,\,4,\,5}\right) = (1)(A),$$

$$\text{where } A = \frac{0,\,1,\,3,\,4,\,7}{0,\,4,\,5} = 0,\,1 \oplus 3\left(\frac{0,\,3,\,4}{0,\,4,\,5}\right) = 0,\,1 \oplus \frac{3}{B},$$

$$\text{where } B = \frac{0,\,4,\,5}{0,\,3,\,4} = 0 \oplus 3\left(\frac{0,\,2}{0,\,3,\,4}\right) = 0 \oplus \frac{3}{C}$$

$$\text{where } C = \frac{0,\,3,\,4}{0,\,2} = 0,\,2 \oplus \frac{3}{0,\,2}$$

Thus the function becomes $(1)\left(0, 1 \oplus \frac{3}{0} \oplus \frac{3}{0, 2} \oplus \frac{3}{0, 2}\right)$. This function is synthesized by starting from the right-hand fraction and working backwards. Fig. 1.6 shows in detail how this is done. The stages in the synthesis, are marked *a*, *b*, . . ., *k*, 1, the arrows showing the signal flow direction in each case. The transfer functions formed at each stage are as follows:

$$a = 2,\ b = \frac{0}{0, 2},\ c = \frac{3}{0, 2},\ d = 2 \oplus \frac{3}{0, 2},\ e = \frac{0}{0, 2} \oplus \frac{3}{0, 2},$$

$$f = \frac{3}{0, 2} \oplus \frac{3}{0, 2},\ g = \frac{0}{0} \oplus \frac{3}{0, 2} \oplus \frac{3}{0, 2},\ h = \frac{3}{0} \oplus \frac{3}{0, 2} \oplus \frac{3}{0, 2},$$

$$= 1 \oplus \frac{3}{0} \oplus \frac{3}{0, 2} \oplus \frac{3}{0, 2},\ k = 0, 1 \oplus \frac{3}{0} \oplus \frac{3}{0, 2} \oplus \frac{3}{0, 2}, \text{ and finally,}$$

$l = (1)\left(0, 1 \oplus \frac{3}{0} \oplus \frac{3}{0, 2} \oplus \frac{3}{0, 2}\right)$, which is the required transfer function.

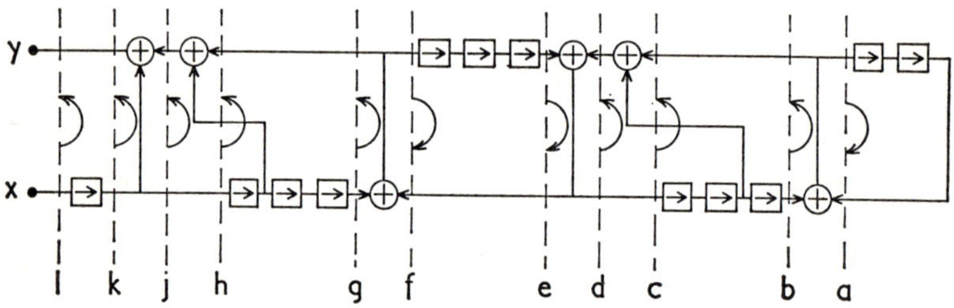

Fig. 1.6. *Continued fraction synthesis of* $\frac{8, 5, 4, 2, 1}{5, 4, 0}$

Notice that this synthesis requires twelve stores, and so twelve initial conditions have to be defined in order to prescribe the value of the output for a given input.

The solution of a difference equation of the form $\phi(D)y = F(D)x$, where $\phi(D)$ is a polynomial of degree m and $F(D)$ is one of degree n, there being no common factors between ϕ and F, requires m or n initial conditions for its solution, whichever is the greater. It must therefore be possible to synthesize the transfer function $F(D)/\phi(D)$ with this number of stores, into which the initial conditions are 'written'. The fourth method, which always leads to a realization using the minimum number of stores, consists of expressing the output y as a nested factorization. Thus the equation $(D^5 \oplus D^4 \oplus 1)y = (D^8 \oplus D^5 \oplus D^4 \oplus D^2 \oplus D)x$ used above is written:

$$y = Dx \oplus D^2x \oplus D^4x \oplus D^5x \oplus D^8x \oplus D^4y \oplus D^5y$$
$$= D(x \oplus D(x \oplus D^2(x \oplus y \oplus D(x \oplus y \oplus D^3x)))),$$

the realization of which is easily seen to be as shown in Fig. 1.7.

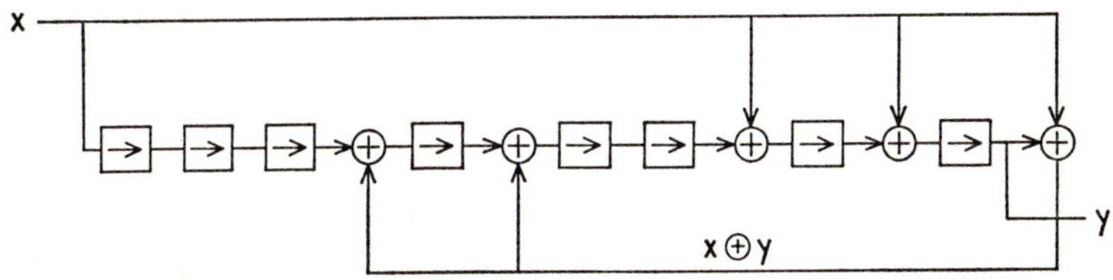

Fig. 1.7. *Synthesis of* $\frac{8, 5, 4, 2, 1}{5, 4, 0}$ *using the least number (eight) of stores*

Another example of this synthesis, in which all three signals x, y, and $x \oplus y$ appear in the factorization, is shown in Fig. 1.8.

The equation in this case is $(D^6 \oplus D^5 \oplus D^2 \oplus D \oplus 1)y = (D^6 \oplus D^4 \oplus D \oplus 1)x$, leading to the nested factorization $y = x \oplus D(x \oplus y \oplus D(y \oplus D^2(x \oplus D(y \oplus D(x \oplus y)))))$.

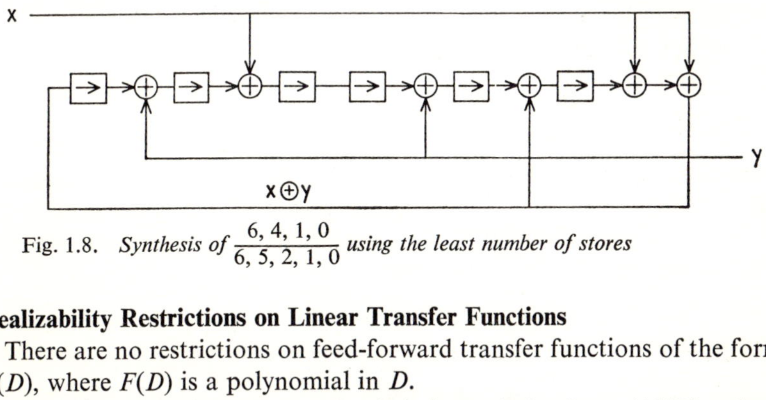

Fig. 1.8. *Synthesis of* $\frac{6, 4, 1, 0}{6, 5, 2, 1, 0}$ *using the least number of stores*

Realizability Restrictions on Linear Transfer Functions

There are no restrictions on feed-forward transfer functions of the form $F(D)$, where $F(D)$ is a polynomial in D.

The only restriction on transfer functions of the form $1/\phi(D)$, where $\phi(D)$ is a polynomial in D, is that $\phi(D)$ should contain the identity operation. In the case of functions of the form $F(D)/\phi(D)$, the denominator must contain the identity operation after removal of any power of D common to F and ϕ. This may be seen by considering, for example, the transfer function $\frac{5, 2, 1}{7, 5, 2}$. Removal of the common factor D leaves the function $\frac{4, 1, 0}{6, 4, 1}$, thus $y_r = y_{r-5} \oplus y_{r-3} \oplus x_{r-3} \oplus x_r \oplus x_{r+1}$, implying that the output y at epoch r depends on the input x at epoch $r + 1$, which is clearly impossible.

For the inverse function $\phi(D)/F(D)$ to be realizable the denominator must as before contain the identity operation after removal of any power D common to F and ϕ, and so if both transfer function and its inverse are to be realizable, the lowest powers of D in F and ϕ must be the same.

Behaviour of Different Realizations of a Given Transfer Function

The various methods of synthesis described above lead in general to realizations having different numbers of stores. It is interesting therefore to compare their actual outputs for given input sequences. If the initial states of all the stores are zero, all realizations give identical outputs for the same input sequence since the transfer functions are the same. Their behaviour differs slightly, however, if non-zero initial conditions are assumed.

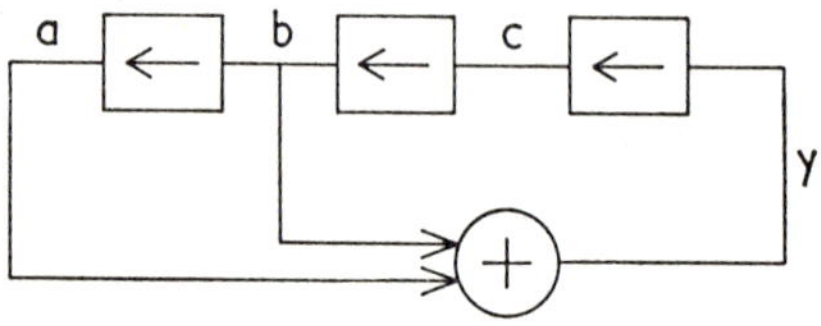

Fig. 1.9. *Circuit satisfying* $y_r \oplus y_{r-2} \oplus y_{r-3} = 0$

Consider the fairly simple transfer function $\dfrac{0, 1, 2}{0, 2, 3}$ in which both numerator and denominator are prime. The circuit satisfying the equation $(D^3 \oplus D^2 \oplus 1)y = 0$, or in recurrence relation form, $y_r \oplus y_{r-2} \oplus y_{r-3} = 0$, is shown in Fig. 1.9. Initial conditions a, b, and c are assumed. The outputs at successive epochs are:

$$y_0 = a \oplus b, \quad y_1 = b \oplus c, \quad y_2 = a \oplus b \oplus c, \quad y_3 = a \oplus c,$$
$$y_4 = a, \quad y_5 = b, \quad y_6 = c, \quad y_7 = a \oplus b, \ldots$$

This sequence has a period of seven and contains all possible combinations of a, b, and c.

Figs. 1.10 (a) and (b) show realizations of the transfer function using cascaded feed-forward and feedback circuits. Initial conditions d and e are assumed in the feed-forward circuit.

The output of circuit (a) at successive epochs is:

$$x_0 \oplus d \oplus e \oplus a \oplus b, \; x_1 \oplus x_0 \oplus d \oplus b \oplus c, \; x_2 \oplus x_1 \oplus d \oplus e \oplus a \oplus b \oplus c,$$
$$x_3 \oplus x_2 \oplus e \oplus a \oplus c, \; x_4 \oplus x_3 \oplus x_0 \oplus e \oplus a, \; x_5 \oplus x_4 \oplus x_1 \oplus d \oplus b,$$
$$x_6 \oplus x_5 \oplus x_2 \oplus x_0 \oplus c, \; x_7 \oplus x_6 \oplus x_3 \oplus x_1 \oplus x_0 \oplus d \oplus e \oplus a \oplus b, \ldots,$$

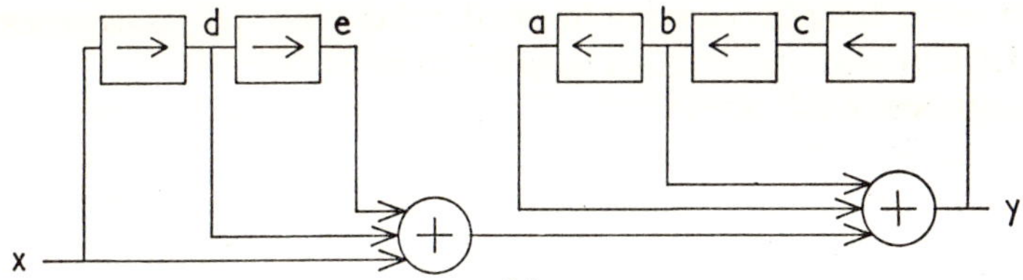

(*a*)

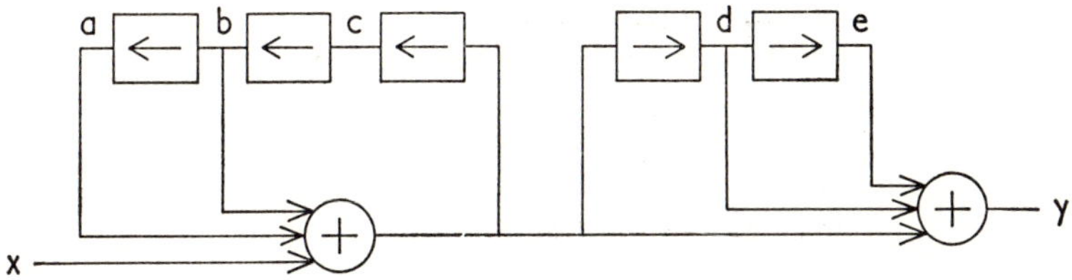

(*b*)

Fig. 1.10. *Cascaded realizations of* $\dfrac{0,\,1,\,2}{0,\,3,\,2}$

and a periodicity of seven is established at once in the initial condition component of the output irrespective of the values of *a*, *b*, *c*, *d*, and *e*. This may be seen by considering the input x to be continuously zero. The output is then simply the sum of the output of the feedback part of the circuit, this output delayed by one epoch, and this output delayed by two epochs.

The corresponding initial condition components in the output of circuit (*b*) are:

$d \oplus e \oplus a \oplus b,\ d \oplus a \oplus c,\ b,\ c,\ a \oplus b,\ b \oplus c,\ a \oplus b \oplus c,\ a \oplus c,\ a,\ b, \ldots$

Again the output repeats itself every seven epochs but the first period begins only after the initial conditions *d* and *e* in the feed-forward part of the circuit have been obliterated, i.e. at the third epoch. It will be shown in the next chapter that if the denominator of the transfer function is not prime, the period of the initial condition component in the two cases is not necessarily the same.

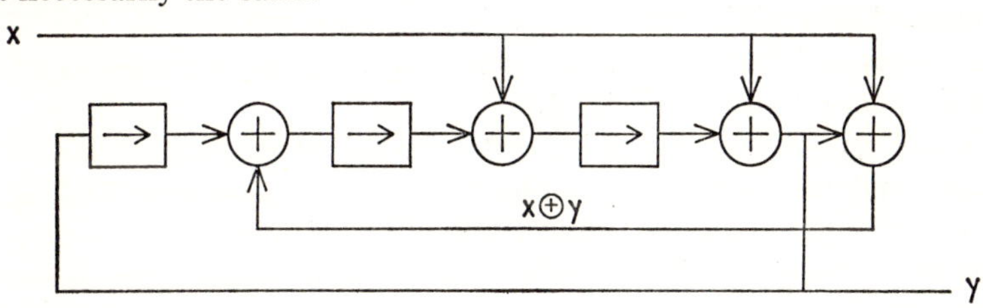

Fig. 1.11. *Realization of* $\dfrac{0,\,1,\,2}{0,\,2,\,3}$ *using nested factorization*

The same transfer function realized using three stores is shown in Fig. 1.11. In this case it is not possible to isolate feed-forward and feedback sections of the circuit and a direct comparison with the realizations shown in Fig. 1.10 is not possible. However, the initial condition components in the output are successively $a, b, a \oplus c, a \oplus b, a \oplus b \oplus c, b \oplus c, c, a, \ldots$ and a periodicity of seven is established at once. This follows from the fact that for zero input the circuit degenerates to that

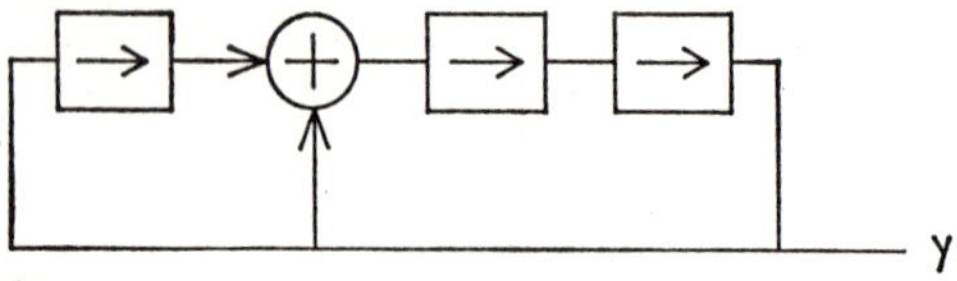

Fig. 1.12. *Alternative circuit satisfying* $y_r \oplus y_{r-2} \oplus y_{r-3} = 0$

shown in Fig. 1.12, which represents an alternative realization of the difference relation $y_r \oplus y_{r-2} \oplus y_{r-3} = 0$. Each store output therefore satisfies this equation.

The partial fraction realization of the transfer function $\frac{0,\,1,\,2}{0,\,2,\,3}$ is shown in Fig. 1.13. The four stores have initial states a, b, c, and d, and the components in the output due to these initial states are successively $a \oplus b \oplus d, c, b \oplus d, a \oplus b \oplus c \oplus d, b \oplus c \oplus d, a \oplus c, a, a \oplus b \oplus d, \ldots$, and once again, though four stores are involved in what is neither strictly

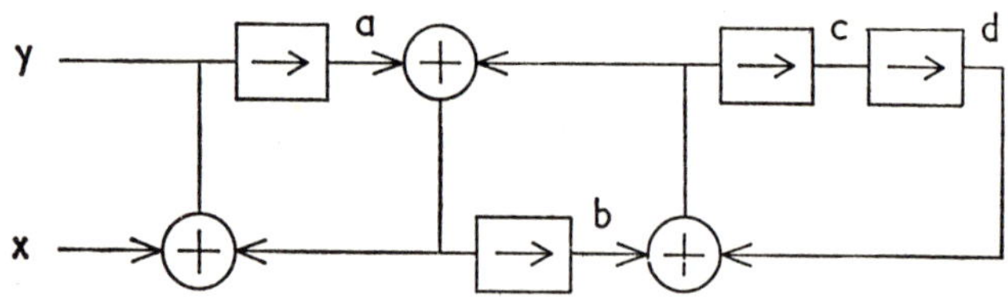

Fig. 1.13. *Partial fraction realization of* $\frac{0,\,1,\,2}{0,\,2,\,3}$

a feed-forward nor a feedback circuit, a periodicity of seven is established at once. Again this is seen by supposing the input x to be zero. One of the four stores and two of the three adders then become redundant and the circuit reduces to that of Fig. 1.12.

It may be noted at this point that a parallel exists between the above circuits and linear systems operating on continuous variables. The output of the latter may be expressed as the sum of a particular integral and a complementary function. For a stable system the complementary function, which depends on initial conditions, dies away leaving only a particular integral determined by the input waveform. Binary systems containing feedback loops are unstable, however, in the sense that they yield an output with no input. The component in their output which depends on initial conditions may therefore be regarded as a complementary function which does not die away, and this component satisfies the equation $\phi(D)y = 0$, where $\phi(D)$ is the denominator of the transfer function.

A special case of interest is the transfer function in which both numerator and denominator polynomials are identical. Figs. 1.14 (*a*) and (*b*) represent two possible realizations of the transfer function $\dfrac{0, 2, 3}{0, 2, 3}$. Initial conditions a, b, c, d, e, and f are assumed.

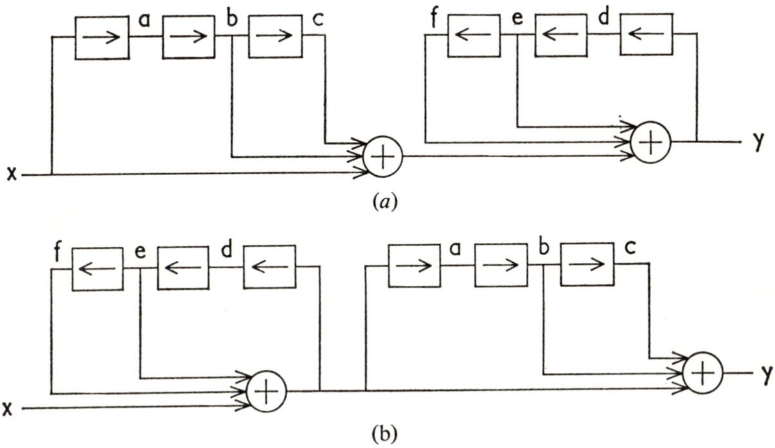

Fig. 1.14. *Alternative realizations of* $\dfrac{0, 2, 3}{0, 2, 3}$

Successive outputs from circuit (*a*) are:

$x_0 \oplus b \oplus c \oplus e \oplus f,\ x_1 \oplus a \oplus b \oplus d \oplus e,\ x_2 \oplus a \oplus b \oplus c \oplus d \oplus e \oplus f,$
$x_3 \oplus a \oplus c \oplus d \oplus f,\ x_4 \oplus c \oplus f, x_5 \oplus b \oplus e, x_6 \oplus a \oplus d,$
$x_7 \oplus b \oplus c \oplus e \oplus f, \ldots$

and from circuit (*b*) are:

$x_0 \oplus b \oplus c \oplus e \oplus f,\ x_1 \oplus a \oplus b \oplus d \oplus e,\ x_2 \oplus a \oplus d,\ x_3, x_4, x_5, \ldots$

The output in the latter case becomes equal to the input after the propagation time through the feed-forward part of the circuit, whereas in the former case the output always contains a component (the 'complementary function') depending on the initial state of all the stores. If the input to circuit (*b*) is always zero, its output necessarily becomes zero after three epochs whatever the initial conditions. Nevertheless the feedback part of the circuit is generating a sequence defined by the equation $(D^3 \oplus D^2 \oplus 1)y = 0$. This sequence is therefore one which produces no output from a feed-forward circuit of transfer function (0, 2, 3), and is therefore referred to as its ‘null sequence’.

Chapter two

The Feedback Shift Register

In chapter one it was shown that a given transfer function could be realized in a number of ways. With an input which was continuously zero it was seen that certain of these realizations, for example the continued fraction and the nested factorization realizations, degenerated into simpler circuits. The cascaded realization in which a feedback circuit representing the denominator of the transfer function supplies a feed-forward circuit representing the numerator (e.g. Fig. 1.10 (*b*)) was seen to yield a sequence ultimately depending only on the input and on the initial conditions of the first (feedback) part of the circuit. The nett output of a cascaded realization of this kind may be regarded as the sum (modulo-two) of phase-shifted versions of the output of the feedback portion of the circuit. Thus in the case quoted, if h_r is the output of the feedback part of the circuit at epoch r, the nett output y_r is $h_r \oplus h_{r-1} \oplus h_{r-2}$, i.e. the sum of h itself, h delayed by one epoch and h delayed by two epochs. The relation between y and h for various feedback and feed-forward circuits cascaded in the above way is discussed in detail later under the heading 'Shift-and-Add Property'. For the present we will be concerned only with the behaviour of straightforward linear feedback circuits of which Figs. 1.4, 1.9, and 1.12 are simple examples. A general form of such circuits is shown in Fig. 2.1. It consists of an n-stage shift register whose input is a feedback digit formed by summing the outputs of the last store and some (or none) of the other stores. The elements $c_1, c_2, \ldots, c_{n-1}$ have transfer function 0 or 1 according to which outputs are fed back. The output of the circuit is normally the

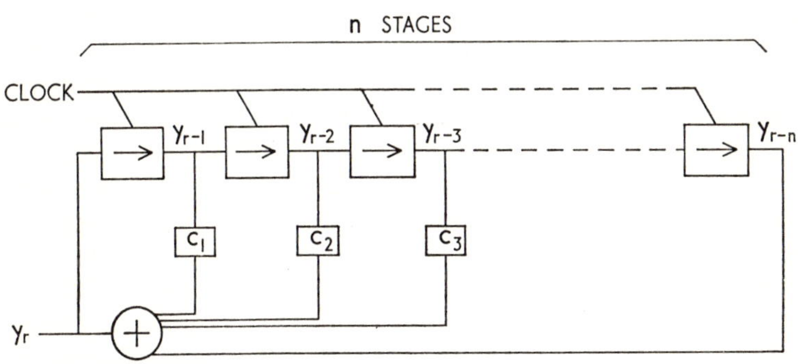

Fig. 2.1. n-*stage shift register with linear feedback*

feedback digit y_r, as shown, and therefore satisfies the recurrence relation defining the feedback digit, namely $y_r \oplus c_1 y_{r-1} \oplus \dots \oplus c_{n-1} y_{r-n+1} \oplus y_{r-n} = 0$. The same equation defines the output from each of the stores, and it may be noted in passing that since the output from any one store is that of its predecessor one epoch earlier, all phase-shifted versions of y up to and including y_{r-n} are available from the register.

Periodicity

Examination of the nth-order recurrence relation above shows that, providing $y_{r-1}, y_{r-2}, \dots, y_{r-n}$ are not all simultaneously zero, the application of successive clock pulses causes y_r to run through a definite sequence of 1s and 0s, the value of y at any epoch r being determined by r and the initial states of the stores, i.e. the values of $y_{r-1}, y_{r-2}, \dots, y_{r-n}$ when $r = 0$. If the initial state of all the stores is zero, however, y remains continuously zero.

The case with $n = 3$, $c_1 = 0$, and $c_3 = 1$, giving the relation $y_r \oplus y_{r-2} \oplus y_{r-3} = 0$, has been considered in chapter one.

A shift register which produces a feedback digit which is a fixed function of the register state (i.e. the states of its stores) is strictly deterministic. Any one state of the register is always followed by the same successive state. Since the number of possible states of the register is finite, and since there is no external input to the register, the feedback digit sequence must ultimately become periodic. The number of possible states of an n-stage register is 2^n, and so this represents the greatest possible periodicity. In particular, if the form of the recurrence relation defining the feedback digit is such that successive clock pulses cause the register to run through all the possible 2^n states, a unique closed cycle of states is produced, irrespective of the initial conditions since these constitute one of the register states. The feedback digit sequence in this case has periodicity 2^n and is called a *full-length sequence*. The linear feedback shift register of Fig. 2.1 can never be made to produce a full-length sequence since the all-zero state gives rise to zero feedback digit. It is possible, however, to choose the feedback coefficients $c_1, c_2, \dots, c_{n-1}$ to make the register run through $2^n - 1$ different states given initial conditions which are not all zero, whereby the feedback digit sequence has a period of $2^n - 1$. In this case the register states form two separate closed cycles, one a degenerate cycle consisting of the all-zero state, and the other a cycle of $2^n - 1$ states. The feedback digit sequence in the first case is a sequence of zeros and in the second case it is a *maximum-length sequence of order* n. The name 'maximum-length sequence' is reserved for a sequence of length $2^n - 1$ produced by a *linear* feedback shift register containing n stores.

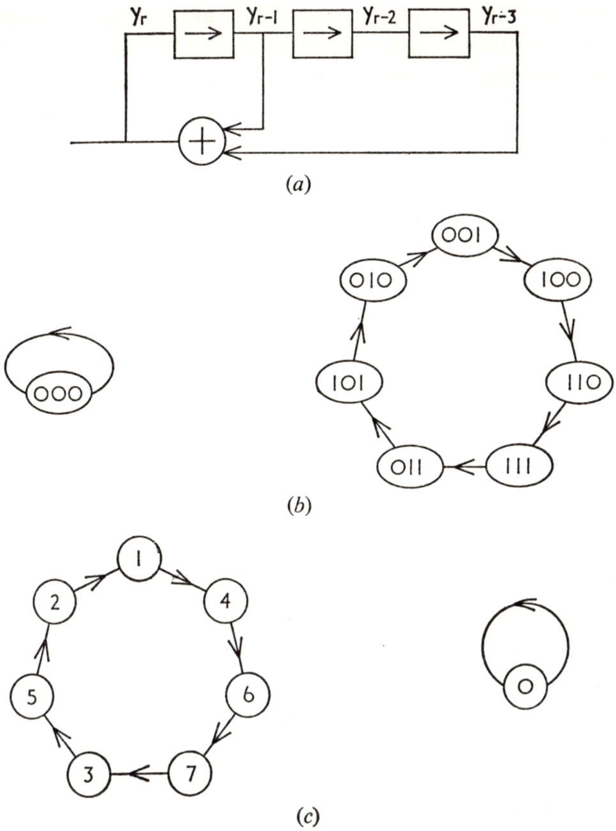

Fig. 2.2. *Cycles from feedback register* (0, 1, 3)

Fig. 2.2 (*a*) shows a feedback register whose output satisfies the equation (0, 1, 3). This equation gives rise to a maximum length sequence of period seven. Fig. 2.2 (*b*) shows the two possible cycles of register states. The digits in the circles represent the states of the stores at each epoch, the right-hand digit being the output of the right-hand store. Fig. 2.2 (*c*) shows an alternative representation which is sometimes used in which the states are represented as decimal numbers. When the register is generating a maximum-length sequence the feedback digit, which is also the output of the circuit, pursues the sequence 1110100 repeats, and it is seen that all the seven non-zero configurations of three binary digits are present in sequence.

In general, feedback shift registers do not produce the greatest possible periodicity. The behaviour of linear feedback registers which do not produce maximum-length sequences is discussed later.

State Transitions—the Complete State Diagram

Consider the general third-order feedback shift register consisting of three binary stores in cascade and feedback logic, as shown in Fig. 2.3. Suppose the states of the stores at a given epoch are *a*, *b*, and *c* (where *a*, *b*, and *c* are either 0 or 1). The next register state is either 0, *a*, *b* if the state *a*, *b*, *c* produces a feedback digit equal to zero, or 1, *a*, *b* if the state *a*, *b*, *c* produces a feedback digit equal to one. Likewise the preceding register state is either *b*, *c*, 0 if the state *b*, *c*, 0 produces a feedback digit equal to *a*,

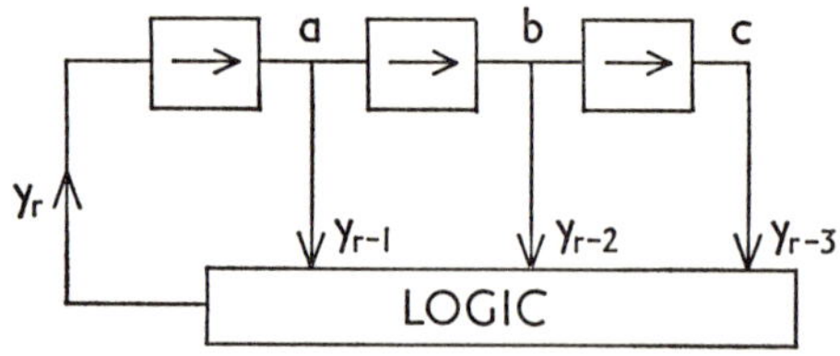

Fig. 2.3. *General three-stage feedback shift register*

or *b*, *c*, 1 if the state *b*, *c*, 1 produces the feedback digit *a*. Fig. 2.4 (*a*) shows the situation. Every register state can therefore lead to one of two states and can have arisen from two states. All five states shown in Fig. 2.4 (*a*) are distinct except when *a*, *b*, and *c* are either all one or all zero, in which case only three of the five states are distinct, or if *a*, *b*, and *c* are alternately one and zero, in which case four of the states are distinct. The possible state transitions in these cases are shown in Fig. 2.4 (*b*).

Although we have considered only a third-order register, the above argument applies to registers of any order.

The complete diagram of possible state transitions ('complete state diagram') of a three-stage shift register is shown in Fig. 2.4 (*c*). Remembering, however, that the circuit of Fig. 2.3 is deterministic, each register state has only one successor, not two, and this constraint limits the number of possible state-transition structures which this circuit can produce. Nevertheless, even in a third-order register 128 different structures are possible. Fig. 2.5 shows some of them, with the feedback logic providing them.

It is seen that confluences of states can occur, implying that states may lie on branches joined to cycles. On the other hand if no such confluences occur, the structure consists of separate cycles. All four structures shown result from non-linear logic, though the logic for the third structure consisting of two separate cycles is quasi-linear.

It will be seen from Fig. 2.4 (*c*) that a number of closed cycles are possible. Expressing the states as decimal numbers the possible cycles are:

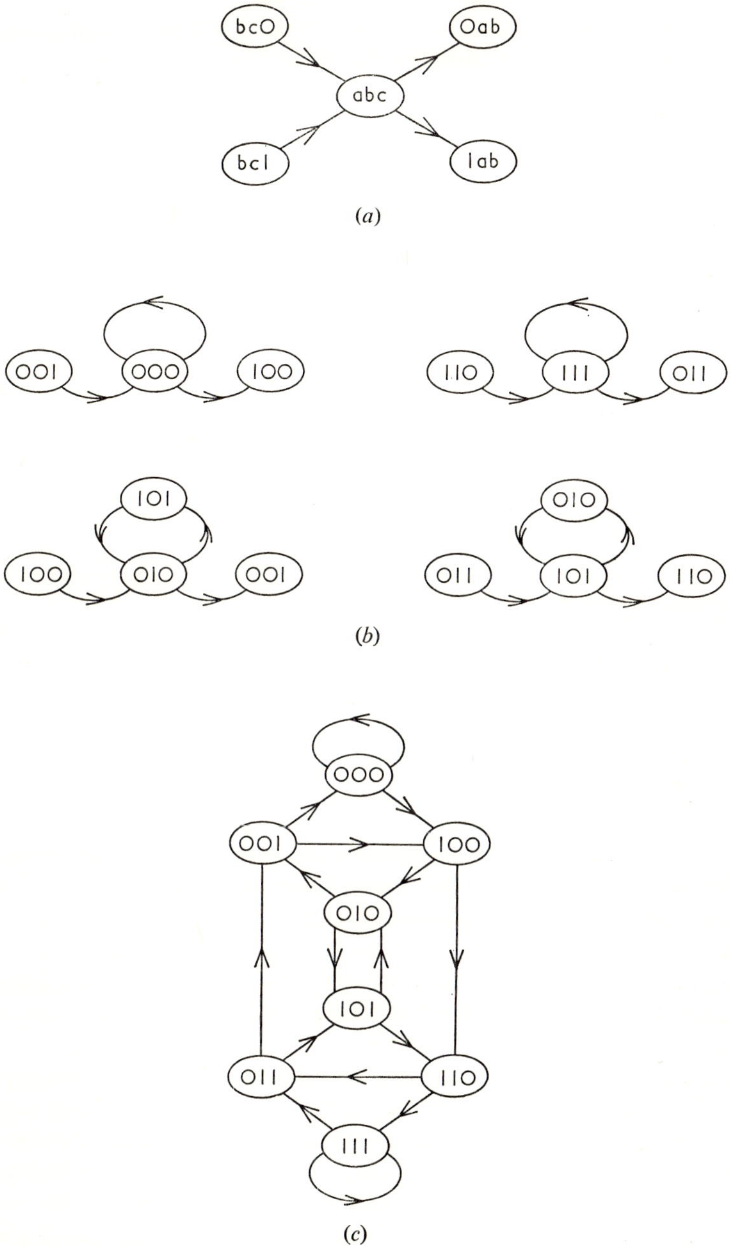

Fig. 2.4. *Possible state transitions of a three-stage register*

0, repeats
7, repeats
2, 5, repeats
1, 4, 2, repeats
3, 5, 6, repeats
0, 4, 2, 1, repeats
3, 5, 6, 7, repeats
1, 4, 6, 3, repeats
0, 4, 6, 3, 1, repeats
1, 4, 6, 7, 3, repeats
0, 4, 6, 7, 3, 1, repeats
1, 4, 2, 5, 6, 3, repeats
1, 4, 6, 3, 5, 2, repeats
0, 4, 2, 5, 6, 3, 1, repeats
0, 4, 6, 3, 5, 2, 1, repeats
1, 4, 6, 7, 3, 5, 2, repeats
1, 4, 2, 5, 6, 7, 3, repeats
0, 4, 2, 5, 6, 7, 3, 1, repeats
0, 4, 6, 7, 3, 5, 2, 1, repeats

There are nineteen cycles in all, and all cycle lengths from one to eight are possible. While no formula has been found giving the number of closed cycles of given lengths in the general nth-order state diagram, it will be shown later that at least one cycle of length $K (1 \leqslant K \leqslant 2^n)$ exists in the diagram.

Condition for Separate Cycles

The condition for no confluences to exist in a state-transition structure of the circuit of Fig. 2.3 is that register states a, b, 0 and a, b, 1 should always yield different feedback digits whatever the (binary) values of a and b. It is then never possible for two states differing in their third digit ($c = y_{r-3}$) to lead to the same successive state. This implies that the variable y_{r-3} must appear as a separate term $\oplus\ y_{r-3}$ in the recurrence relation for y_r.

What has been said in respect of a circuit containing three stores applies equally to one containing n stores except that each register state is now described by n digits, and so for an nth-order state-transition structure to consist of one or more separate cycles, y_{r-n} must appear as a separate term $\oplus y_{r-n}$ in the recurrence relation defining the feedback digit y_r. The linear feedback register of Fig. 2.1 clearly satisfies this condition, so all the possible 2^n states of this register lie on (two or more) closed unconnected cycles.

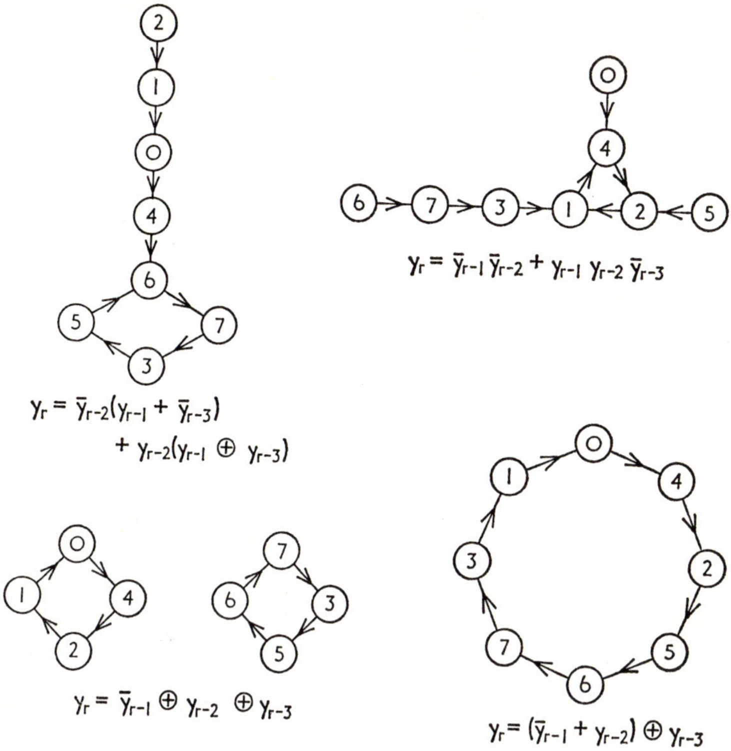

Fig. 2.5. *Some cycle structures of a three-stage register, with corresponding feedback logic*

A consequence of the above condition is that if a truth table is drawn up for the feedback digit y_r in terms of the store states $y_{r-1}, \ldots, y_{r-n}$, in which the first column is the digit y_{r-n}, each digit in the top half of the y_r column is paired with its inverse in the corresponding position in the lower half of the column.

It is clear that if an nth-order state-transition structure consists of separate cycles, there being K_1 cycles of length one ($K_1 = 0$, 1 or 2), K_2 of length two ($K_2 = 0$ or 1), K_3 of length three, . . . $K_{(2^n)}$ of length $2^n (K_{(2^n)} = 0$ or 1).

$K_1 + 2K_2 + 3K_3 + \ldots + 2^n K_{(2^n)} = 2^n =$ total number of register states.

Chapter three

The Cycling Register

The simplest feedback register is the cycling or circulating register. This consists of a number of binary stores in cascade, the input to the first store (the feedback digit) being the output of the last store. Before considering the behaviour of the cycling register, some of the properties of the arithmetic functions of Mobius and Euler will be outlined.

The Mobius and Euler Functions

The Mobius and Euler functions are central to the study of recurring sequences. The Mobius function arises when calculating the inverse $f(n)$ of a function $F(n)$ given by

$$F(n) = \sum_{d|n} f(d),$$

where n is an integer and d is a divisor of n. The number one is regarded as a divisor of all n. The following table gives the first few values of F in terms of f:

n	d	$F(n)$
1	1	$f(1)$
2	1, 2	$f(1)+f(2)$
3	1, 3	$f(1)+f(3)$
4	1, 2, 4	$f(1)+f(2)+f(4)$
5	1, 5	$f(1)+f(5)$
6	1, 2, 3, 6	$f(1)+f(2)+f(3)+f(6)$
7	1, 7	$f(1)+f(7)$
8	1, 2, 4, 8	$f(1)+f(2)+f(4)+f(8)$

Solving for $f(n)$ in each case we obtain:

$$\begin{aligned}
f(1) &= F(1)\\
f(2) &= F(2) - F(1)\\
f(3) &= F(3) - F(1)\\
f(4) &= F(4) - F(2)\\
f(5) &= F(5) - F(1)\\
f(6) &= F(6) - F(3) - F(2) + F(1)\\
f(7) &= F(7) - F(1)\\
f(8) &= F(8) - F(4)
\end{aligned}$$

and generally, $f(n) = \sum_{d|n} \mu(n/d)F(d) = \sum_{d|n} \mu(d)F(n/d)$,

where μ is the Mobius function, defined as follows:

$\mu(n) = 1$ if $n = 1$,
$= 0$ if n contains any prime factor raised to a power greater than one,
$= (-1)^k$ if n is the product of k prime factors none of which is raised to a power greater than one.

The formula for $f(n)$ is Mobius' Inversion Formula. A formal proof may be found in any text on Number Theory.* The converse of the above relation is also true, namely if $f(n)$ is defined by $\sum_{d|n} \mu(n/d)F(d)$, then $F(n) = \sum_{d|n} f(d)$. An important corollary of this formula follows by considering the inverse $f(n)$ of a function $F(n)$ given by $F(n) = \prod_{d|n} f(d)$, i.e. where $F(n)$ is the product rather than the sum of the values of some other function taken over the divisors of n.

Taking logarithms, $\log F(n) = \log \prod_{d|n} f(d) = \sum_{d|n} \log f(d)$.

$$\begin{aligned} \text{Therefore} \quad \log f(n) &= \sum_{d|n} \mu(d) \log F(n/d), \text{ by the inversion formula,} \\ &= \sum_{d|n} \log [F(n/d)^{\mu(d)}] \\ &= \log \left\{ \prod_{d|n} F(n/d)^{\mu(d)} \right\}, \end{aligned}$$

$$\text{and so} \qquad f(n) = \prod_{d|n} F(n/d)^{\mu(d)} = \prod_{d|n} F(d)^{\mu(n/d)}.$$

This formula can be verified in the same way as the summation case above. Its converse is true, as in the summation case.

The Euler function $\phi(n)$ gives the number of numbers less than an integer n and prime to it. The number one is taken as prime to all n. Thus $\phi(1) = 1, \phi(2) = 1, \phi(3) = 2, \phi(4) = 2, \phi(5) = 4, \phi(6) = 2$, etc. If numbers $a, b, c, \ldots$, are prime to each other, $\phi(abc\ldots) = \phi(a) . \phi(b) . \phi(c) \ldots$ from which it follows† that if $n = p^\alpha q^\beta r^\gamma \ldots$, where $p, q, r, \ldots$, are prime, $\phi(n) = n(1 - 1/p)(1 - 1/q)(1 - 1/r) \ldots$ An important property of the Euler function, which follows from this, is the relation $n = \sum_{d|n} \phi(d)$. Applying the Mobius inversion formula to this relation gives $\phi(n) = \sum_{d|n} \mu(d)(n/d)$, or $\phi(n)/n = \sum_{d|n} \mu(d)/d$.

* Hardy and Wright, 'Introduction to the Theory of Numbers'.
† Hall and Knight, 'Higher Algebra'.

Other interesting relations arising from the Mobius inversion formula, are, for example,

(i) $\sum_{d/n} \mu(d) = 0$ (except for $n = 1$)

(ii) $\sum_{d/n} \mu(d)d(n/d) = 1$, where $d(n)$ is the number, $\sum_{d/n} 1$, of divisors of n

(iii) $\lambda(n) = \sum_{d/n} \mu(n/d) \log d$, where

$\lambda(n) = \log p$ if n is a power of a single prime, p,
$= 0$ if n contains different primes.

The function $\lambda(n)$ is important in the theory of prime numbers.

Number of Cycles Produced by a Cycling Register

The nth-order cycling register, as shown in Fig. 3.1, consists of n stores in cascade, the input to the first being the output of the nth store. The feed-

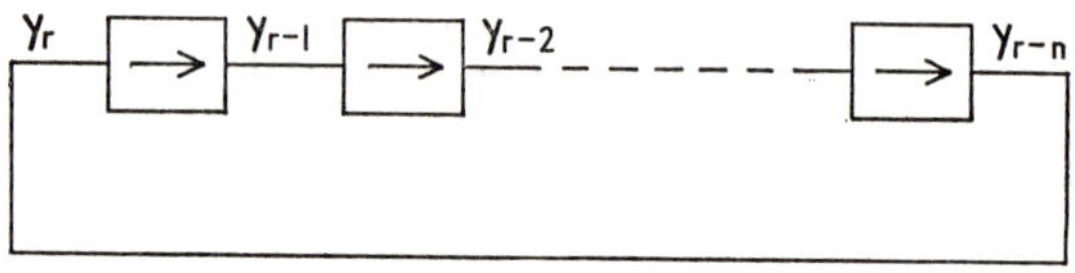

Fig. 3.1. *Cycling register of order n*

back digit y_r is defined by the equation $y_r = y_{r-n}$, written more briefly $(0, n)$. Since the term y_{r-n} appears separately the state-transition structure of all cycling registers consists of closed unconnected cycles.

Evidently an n-stage cycling register is capable of producing all possible sequences of period n, as well as all those whose period divides n, since any register state whatever consisting of n digits can be set up initially. Each of these sequences corresponds to a closed cycle of states, the length of each cycle being the period of the sequence. Two such cycles, common to all cycling registers, consist of the all-zero and the all-one states. These are each of length one and the sequences corresponding to them are 000 . . . and 111 . . . respectively. In the case of $n = 4$, for example, six sequences are possible, namely 111 . . . (period one), 000 . . . (period one), 0101 . . . (period two), 10001000 . . . (period four), 11001100 . . . (period four), and 11101110 . . . (period four). These six sequences correspond to the six cycles shown in Fig. 3.2.

Obviously no other sequences of periods one, two, or four could be generated by any other cycling register, or indeed by any other feedback shift register, although the states themselves in the case of a register of

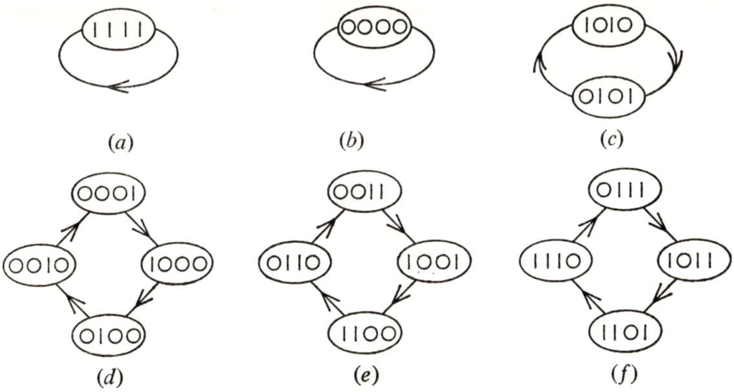

Fig. 3.2. *Cycle structure of fourth-order cycling register*

order other than four would contain a different number of digits. For example the sequence 0101 . . . generated by a three-stage register would correspond to the cycle

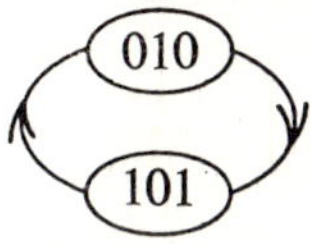

The number $\nu(n)$ of possible sequences of period n, i.e. the number of possible cycles of length n, may be found by supposing them to be produced by an n-stage cycling register. As we have seen, this register also produces all sequences whose periods divide n, and so the number 2^n of n-digit states equals $\sum_{d|n} d\nu(d)$, since a sequence of period d accounts for d register states. Thus

$$\nu(n) = \frac{1}{n}\sum_{d|n} \mu\left(\frac{n}{d}\right)2^d = \frac{1}{n}\sum_{d|n} \mu(d)2^{n/d},$$

by Mobius' inversion theorem.

This formula shows that there are two sequences of period one, one of period two, two of period three (011011 . . ., and 001001 . . .), three of period four, etc. It may be noted that if n is prime, two register states lie on cycles of length one and the other $2^n - 2$ states lie on cycles of length n. Therefore n divides $2^n - 2$ (Fermat's theorem).

The sequence corresponding to *any* cycle of length K obeys the recurrence relation $y_r = y_{r-k}$ and may therefore be generated by a K-stage cycling register. It follows that $\nu(K)$ is the upper limit to the number of cycles of length K which can be generated by *any* feedback shift register.

The total number of cycles $Z(n)$ produced by an nth-order cycling register is the sum of the number of cycles of length equal to each divisor of n, i.e.

$\sum_{d/n} \nu(d)$. Using the formula for $\nu(n)$ given above, it may easily be shown that $Z(n)$ also equals $\frac{1}{n}\sum_{d/n} \phi(d)2^{n/d}$. For n odd, $Z(n)$ is necessarily even since it is the sum of even terms. Since $\phi(2^k \times$ an odd number greater than one) contains the terms $2^k(1 - \frac{1}{2})$, and at least one other even factor, it is a multiple of 2^k. It follows that for $n = 2^k \times$ an odd number greater than one, $Z(n)$ is again the sum of even terms and is therefore even. For the same reason, if $n = 2^k$, $Z(n)$ is once more the sum of even terms and is therefore even except for $k = 1$, i.e. for $n = 2$. Thus $Z(n)$ is even for all n except $n = 2$, for which $Z(2) = 3$. This fact has great relevance in the theory of non-linear shift register sequences considered later. The first few values of $Z(n)$ are:

$$Z(1) = 2,\ Z(2) = 3,\ Z(3) = 4,\ Z(4) = 6,\ Z(5) = 8,\ Z(6) = 14, \text{ etc.} \ldots$$

It is believed, though it has not been proved, that of all feedback shift registers of order n the cycling register produces the greatest number of cycles, i.e. $Z(n)$ is believed to be the greatest number of cycles derivable from an nth-order register. It is worth noting, however, that the number of closed cycles on the complete state diagram of a general nth-order feedback register greatly exceeds $Z(n)$. Thus for $n = 3$, the number of cycles on the state diagram was seen to be nineteen, whereas $Z(3)$ is only four.

Successive States of a Cycling Register

If successive states of an nth-order cycling register are expressed as numbers $s_r, s_{r+1}, s_{r+2} \ldots$, the state s_r is given in terms of the succeeding state s_{r+1} by the relation $s_r = 2s_{r+1} \pmod{2^n - 1}$. The states are given in the reverse order by this equation because we have regarded store states as propagating to the right whereas the digit significance in a state increases to the left. Suppose the state s_{r+1} is $(abc \ldots K)$, where the n digits $a, b, c, \ldots, K$ are binary digits. Doubling s_{r+1} produces the number $(bc \ldots KO) + 2^n a$. This number $\pmod{2^n - 1}$ is the number $(bc \ldots Ka)$, which is the preceding state s_r. As an example, the states on one of the cycles produced by a fifth-order cycling register read in reverse order are: 00111, 01110, 11100, 11001, 10011. Expressed as decimal numbers these are: 7, 14, 28, 25 = 56 (mod 31), 19 = 50 (mod 31). Notice that 7 = 38 (mod 31).

Chapter four

Sequences from Linear Feedback Shift Registers

The general linear feedback shift register consists of a number of binary stores in cascade, the input to the first of which is the modulo-two sum of the outputs of the last store and intermediate stores. The digit y_r fed back at epoch r is therefore given by the recurrence relation $y_r = c_1 y_{r-1} \oplus c_2 y_{r-2} \oplus \ldots \oplus c_{n-1} y_{r-n+1} \oplus y_{r-n}$, where the coefficient c_i is one if the output of the ith store contributes to the feedback digit and is zero if it does not.

It is convenient, when dealing with the linear registers, to adopt the following notation:

1. A recurrence relation $y_r \oplus c_1 y_{r-1} \oplus \ldots \oplus c_{n-1} y_{r-n+1} \oplus y_{r-n} = (1 \oplus c_1 D \oplus \ldots \oplus c_{n-1} D^{n-1} \oplus D^n) y_r = 0$ will be referred to as the *equation* $(0, c_1, 2c_2, \ldots (n-1)c_{n-1}, n)$. It is always understood, therefore, that the expression in the bracket is zero. For example, the recurrence relation $y_r \oplus y_{r-1} \oplus y_{r-4} = 0$ would be referred to as the equation (0, 1, 4). The equation (0, 1, 4) clearly implies equations (1, 2, 5), (2, 3, 6), etc.

2. The *left-hand side* of the nth-order recurrence relation in Section 1 above will be referred to as the *polynomial* $(0, c_1, 2c_2, \ldots (n-1)c_{n-1}, n)$. Thus the left-hand side of the equation $y_r \oplus y_{r-1} \oplus y_{r-4} = 0$ would be referred to as the polynomial (0, 1, 4).

In this chapter we consider the feedback digit sequences generated by various classes of linear recurrence relations of the above form, or what amounts to the same thing, since the equations are linear, the cycle structures associated with the corresponding linear feedback shift registers.

Periods of the Sequences Defined by a Linear Recurrence Relation

Sequences produced by the equation $(0, s)$, i.e. by a cycling register consisting of s stages, have period s or a submultiple of s.

Consider the equation (0, 1, 3). If we divide the polynomial (0, 1, 3) into the binomial $(0, s)$ as follows:

```
        0124
013 | 0     s
      013
      ---
       13
       124
       ---
        234
        235
        ---
         45s
         457
         ---
```

we find that the least value of s for which (0, 1, 3) divides (0, s) exactly is seven.

It will be seen that the process of division amounts to subtracting from (0, s) the polynomials (0, 1, 3), (1, 2, 4), (2, 3, 5), and (4, 5, 7), all of which are zero. The equation (0, 1, 3) therefore implies that the binomial (0, s) is also zero when $s = 7$. Consequently the period of at least one of the sequences generated by the equation (0, 1, 3) must be seven. Other sequences may occur having different periods, but their periods must divide seven. The possible sequences in this instance are 1110100, repeats, and 0, repeats (of period one, which is a factor of seven). The cycle structure therefore consists of two cycles, one comprising seven non-zero register states and the other the all-zero state (see Fig. 2.2 (*b*)). The number seven is referred to as the *exponent* of the polynomial (0, 1, 3).

Since the major cycle contains all the possible (in this case $2^3 - 1$) non-zero register states the sequence is called a maximum-length or *m*-sequence, and its period is $2^3 - 1$. The polynomial (0, 1, 3) itself is called a maximal polynomial.

Consider now the equation (0, 1, 4). The least value of s for which the polynomial (0, 1, 4) divides (0, s) exactly is fifteen. This is therefore the exponent of the polynomial (0, 1, 4), and so at least one of the sequences produced by the equation (0, 1, 4) has period fifteen. Others of different period may exist, but their periods must divide fifteen. We notice that $15 = 2^4 - 1$, so in fact the polynomial is maximal. Only one non-zero sequence (of period 15) therefore exists, and the major cycle consists of all fifteen non-zero states of a four-stage register. The number 15, however, has factors 3 and 5. It will be recalled that a fifteen-stage cycling register is capable of producing sequences of period 15, 5, and 3, so the question arises as to why in the present case periods 3 and 5 are not possible. If we divide the polynomial (0, 5) by the polynomial (0, 1, 4) as follows:

```
014 | 0 5
      014
      ---
       145
       125
       ---
        24
```

we get the remainder (2, 4), which is equivalent to (0, 2). Since polynomials (0, 1, 4) and (0, 2) have no common factors, no sequence other than the all-zero sequence can make them simultaneously zero, and consequently the equations (0, 1, 4) and (0, 5) are incompatible. Likewise, dividing (0, 1, 4) by (0, 3) gives the remainder (3), which is equivalent to the monomial (0). Again, only the all-zero sequence makes both the nomials (0, 3) and (0) simultaneously zero.

Consider now the equation (0, 1, 5). The least value of s for which polynomial (0, 1, 5) divides binomial (0, s) exactly is 21, which is therefore the exponent of (0, 1, 5). The number 21 has factors 3 and 7 and the division process shows the circumstances in which this equation can yield periods 3 and 7. Dividing first (0, 1, 5) into (0, 7) as follows:

```
      012
015 | 0
      015
      ---
       15
      126
      ---
       2567
        237
        ---
        356
```

we obtain the remainder (3, 5, 6) which is equivalent to (0, 2, 3). The polynomial (0, 2, 3) is a factor of the polynomial (0, 1, 5), so if the initial conditions of the corresponding five-stage register are chosen to satisfy the equation (0, 2, 3), for example 11001, both the equations (0, 1, 5) and (0, 2, 3) are satisfied. The equation (0, 7) is therefore also satisfied and the sequence generated, namely 1100101, repeats, has period seven. (Note, in writing down a sequence, successive terms are $y_{r-4}, y_{r-3}, y_{r-2}, y_{r-1}, y_r, \dots$ and so equation (0, 2, 3) generates a sequence in which any term is the modulo-two sum of the last-but-two and the last-but-three terms.)

Dividing polynomial (0, 1, 5) by binomial (0, 3) yields the remainder (3, 4, 5), which is equivalent to (0, 1, 2). This is also a factor of (0, 1, 5), so by choosing the initial conditions to satisfy equation (0, 1, 2) eq. 11011, both equations (0, 1, 5) and (0, 1, 2) are satisfied, and so also therefore is equation (0, 3). The sequence generated thus has period 3 (110, repeats).

The cycle structure for equation (0, 1, 5) therefore comprises four cycles, one consisting of the all-zero state, another of three states, another of seven states, and another consisting of the remaining 21 states. All the possible 32 states of the five-stage register are therefore accounted for.

In general, if an nth-order polynomial is found to have an exponent $2^n - 1$ it is certain that the polynomial is prime and maximal, and the cycle structure comprises the all-zero cycle of length one and another cycle consisting of all the other $2^n - 1$ register states. It is never possible for a factorable polynomial to be maximal since the equation corresponding to such a polynomial is also satisfied by sequences which satisfy the equations corresponding to its factors. Consequently the cycle structure must contain more than one cycle of non-zero states.

Consider now the equation (0, 3, 6). Division shows that the polynomial (0, 3, 6) has exponent 9, and since the polynomial is prime, nine is the only possible period of the sequences produced by this equation. A six-stage register has $2^6 - 1 = 63$ possible non-zero states, and so 63 is the period of a sixth-order m-sequence. The polynomial (0, 3, 6) is therefore a *non-maximal* prime polynomial, and the cycle structure associated with it comprises the all-zero state cycle and seven cycles each of length nine. This example shows that one can expect to find non-maximal prime polynomials of order n whenever $2^n - 1$ is not prime, and conversely a prime polynomial of order n is necessarily maximal if $2^n - 1$ is prime since in this case it is impossible to subdivide the $2^n - 1$ non-zero register states into cycles of equal length less than $2^n - 1$.

Since the number $2^n - 1$ and any factor thereof is odd, both the number and the length of non-zero cycles associated with either maximal or non-maximal prime polynomials are odd.

Subdivision of the non-zero register states into a number of cycles of equal length is not peculiar to non-maximal prime polynomials. For example the equation (0, 1, 2, 3, 4, 5, 6) yields nine cycles all of length seven, even though the polynomial (0, 1, 2, 3, 4, 5, 6) factorizes into (0, 1, 3) (0, 2, 3).

Significance of Quotient

In the division of (0, 7) by (0, 1, 3) shown above the quotient (0, 1, 2, 4), regarded as an equation, evidently also produces linear sequences of period seven (and one) since (0, 1, 2, 4) divides (0, 7) to yield (0, 1, 3). It has an additional significance, however.

Suppose a polynomial is divided into the monomial (0) and a '1' is written for each term of the quotient and a '0' for each space between the terms the resulting sequence satisfies the equation corresponding to the polynomial. Thus the quotient formed by dividing (0, 1, 3) into (0) is (0, 1, 2, 4, 7, 8, 9, 11, . . .) and the corresponding sequence is therefore 111010011101 . . ., which is the non-zero sequence produced by equation (0, 1, 3). The proof is as follows: Suppose the divisor is the polynomial $1 \oplus a_1D \oplus a_2D^2 \oplus \ldots \oplus D^n$ and the quotient is the polynomial $b_0 \oplus b_1D \oplus b_2D^2 \oplus \ldots$ The product of these must be identically equal to one, so $b_0 = 1$ and the coefficients of D, D^2, etc., in the product must all be zero.

Therefore
$$b_1 \oplus a_1b_0 = 0 \ldots (1)$$
$$b_2 \oplus a_1b_1 \oplus a_2b_0 = 0 \ldots (2)$$
$$- - - - - - - - - - - - - - -$$
$$b_n \oplus a_1b_{n-1} \oplus \ldots \oplus a_{n-1}b_1 \oplus b_0 = 0 \ldots (n)$$
$$b_{n+1} \oplus a_1b_n \oplus \ldots \oplus a_{n-1}b_2 \oplus b_1 = 0 \ldots (n+1)$$
$$- - - - - - - - - - - - - - -$$

If we now regard $b_0, b_1, b_2, \ldots$ as terms of a sequence, where each are 0 or 1, equations (n), $(n + 1)$, etc., all define a sequence b_r satisfying the equation $(1 \oplus a_1D \oplus \ldots \oplus a_{n-1}D^{n-1} \oplus D^n)b_r = 0$, which is the equation corresponding to the dividing polynomial.

As an alternative to the division procedure equations (1) to $(n - 1)$ may be used to calculate the other $n - 1$ initial terms b_1 to b_{n-1} of the sequence. Thus $b_1 = a_1$, $b_2 = a_1 \oplus a_2$, $b_3 = a_1 \oplus a_3$, $b_4 = a_1 \oplus a_2 \oplus a_1a_2 \oplus a_4$, $b_5 = a_1 \oplus a_1a_2 \oplus a_1a_3 \oplus a_5$, etc. . . .

Reverse Sequences

If the equation $y_r \oplus c_1y_{r-1} \oplus \ldots \oplus c_{n-1}y_{r-n+1} \oplus y_{r-n} = 0$ generates a certain set of sequences, the equation $y_r \oplus c_1y_{r+1} \oplus \ldots \oplus c_{n-1}y_{r+n-1} \oplus y_{r+n} = y_r \oplus c_{n-1}y_{r-1} \oplus \ldots \oplus c_1y_{r-n+1} \oplus y_{r-n} = 0$ generates the same sequences in reverse. Any non-maximum length sequence generated by the first equation is generated in reverse by the second equation if the initial conditions are also reversed. For example the equations (0, 1, 4) and (0, 3, 4) generate reverse sequences, and the polynomials (0, 1, 4) and (0, 3, 4) are said to be conjugate.

Dual Sequences

A sequence $\{z_r\}$ is said to be the dual of sequence $\{y_r\}$ if the digits of $\{z_r\}$ are the complements of the corresponding digits of $\{y_r\}$. Sequence $\{z_r\}$

may therefore be produced by the termwise addition of the all-ones sequence generated by equation (0, 1) to $\{y_r\}$, and so the equation producing $\{z_r\}$ is (0, 1) times the equation producing $\{y_r\}$.

Cycle Structures Associated with Factorable Polynomials

A convenient way of obtaining the cycle structure associated with a factorable polynomial is to construct a combination of feedback registers, each of which satisfies the equation corresponding to one factor, in such a way that the combined circuit satisfies the equation corresponding to the original polynomial.

Suppose Z_r is defined by the equation $\phi(D)Z_r = 0$. If $\phi(D)$ can be factorized into relatively prime factors $\phi_1(D)$, $\phi_2(D)$, $\phi_3(D)$, . . ., the arrangement shown in Fig. 4.1 produces an output Z_r satisfying $\phi(D)Z_r = 0$, where ϕ_1, ϕ_2, ϕ_3, . . ., are feedback shift registers satisfying equations $\phi_1(D)Z_{1r} = 0$, $\phi_2(D)Z_{2r} = 0$, $\phi_3(D)Z_{3r} = 0$, etc. . . . This follows because

$$\begin{aligned}\phi(D)Z_r &= [\phi_1(D)\phi_2(D)\phi_3(D)\ldots]Z_r\\ &= [\phi_1(D)\phi_2(D)\phi_3(D)\ldots](Z_{1r} \oplus Z_{2r} \oplus Z_{3r} \oplus \ldots)\\ &= [\phi_2(D)\phi_3(D)\ldots]\phi_1(D)Z_{1r}\\ &\oplus [\phi_1(D)\phi_3(D)\ldots]\phi_2(D)Z_{2r}\\ &\oplus [\phi_1(D)\phi_2(D)\ldots]\phi_3(D)Z_{3r}\\ &= 0, \text{ as required}\end{aligned}$$

The arrangement of Fig. 4.1 does not hold if any two or more of the polynomials $\phi_1(D)$, $\phi_2(D)$, $\phi_3(D)$ . . . have factors in common. Suppose, for example, $\phi_1(D) = f_0(D)f_1(D)$, $\phi_2(D) = f_0(D)f_2(D)$, and $\phi_3(D) = f_3(D)$. We have $Z_r = Z_{1r} \oplus Z_{2r} \oplus Z_{3r}$, and therefore

$$\begin{aligned}&[f_0(D)f_1(D)f_2(D)f_3(D)]Z_r = [f_2(D)f_3(D)][f_0(D)f_1(D)]Z_{1r}\\ &\oplus [f_1(D)f_3(D)][f_0(D)f_2(D)]Z_{2r}\\ &\oplus [f_0(D)f_1(D)f_2(D)]f_3(D)Z_{3r} = 0\end{aligned}$$

So Z_r satisfies the equation $f_0(D)f_1(D)f_2(D)f_3(D)Z_r = 0$, which is of lower order than the equation $\phi_1(D)\phi_2(D)\phi_3(D)Z_r = 0$.

Polynomial Consisting of Unrepeated Prime Factors

In this case $\phi_1(D)$, $\phi_2(D)$, $\phi_3(D)$, etc., in Fig. 4.1 are themselves prime. The sequences which Z_r may pursue comprise the all-zero sequence, when the initial conditions of ϕ_1, ϕ_2, ϕ_3, etc. are all zero, sequences produced by each register operating independently, i.e. when the initial conditions of all but one register are zero, and sequences formed by the termwise addition of these taken in pairs, in threes, etc., i.e. when the initial conditions of all but two, all but three, etc., registers are zero. Since the sequences produced by the registers themselves are all different no two of the above sequences are the same. If any of the factors are non-maximal the corresponding register produces a number of different sequences (all of the same period), and each of these has to be considered when enumerating all possible sequences.

The period of any one sequence is the lowest common multiple of the periods of its constituent sequences. Thus, for example, termwise addition of a sequence of period three to one of period seven yields one sequence of period 21, irrespective of the relative phase of the sequences, since the numbers 3 and 7 are relatively prime. The corresponding cycle consists therefore of 21 states. On the other hand termwise addition of a sequence of period 15 to one of period 63 yields one of three possible sequences each of period 315, and in this case the $15 \times 63 = 945$ states form three cycles each consisting of 315 states. The three different sequences are obtained by adding one of the constituent sequences to three successively phase-shifted versions of the other as follows:

$$
\begin{array}{ll}
\text{(i)} & x_1\, x_2\, x_3 \cdots\cdots\cdots\cdots\cdots\cdots x_{63} x_1 \\
 & y_1\, y_2\, y_3 \cdots\cdots y_{15} y_1 \cdots\cdots y_3\, y_4 \\
\hline
 & x_1\, x_2\, x_3 \cdots\cdots\cdots\cdots\cdots\cdots\cdots x_1 \\
 & \oplus\ \oplus\ \oplus \qquad\qquad\qquad\qquad\qquad \oplus \\
 & y_1\, y_2\, y_3 \cdots\cdots\cdots\cdots\cdots\cdots\cdots y_4
\end{array}
$$

$$
\begin{array}{ll}
\text{(ii)} & x_1\, x_2\, x_3 \cdots\cdots\cdots\cdots\cdots\cdots x_{63} x_1 \\
 & y_{15} y_1\, y_2 \cdots\cdots y_{15} y_1 \cdots\cdots y_2\, y_3 \\
\hline
 & x_1\, x_2\, x_3 \cdots\cdots\cdots\cdots\cdots\cdots\cdots x_1 \\
 & \oplus\ \oplus\ \oplus \qquad\qquad\qquad\qquad\qquad \oplus \\
 & y_{15} y_1\, y_2 \cdots\cdots\cdots\cdots\cdots\cdots\cdots y_3
\end{array}
$$

$$
\begin{array}{ll}
\text{(iii)} & x_1\, x_2\, x_3 \cdots\cdots\cdots\cdots\cdots\cdots x_{63} x_1 \\
 & y_{14} y_{15} y_1 \cdots\cdots y_{15} y_1 \cdots\cdots y_1\, y_2 \\
\hline
 & x_1\, x_2\, x_3 \cdots\cdots\cdots\cdots\cdots\cdots\cdots x_1 \\
 & \oplus\ \oplus\ \oplus \qquad\qquad\qquad\qquad\qquad \oplus \\
 & y_{14} y_{15} y_1 \cdots\cdots\cdots\cdots\cdots\cdots\cdots y_2
\end{array}
$$

An explicit formula for the number and length of cycles in the case of a general polynomial of this kind is given by Elspas,[1] but in fact they can readily be worked out in the above manner for any specific case.

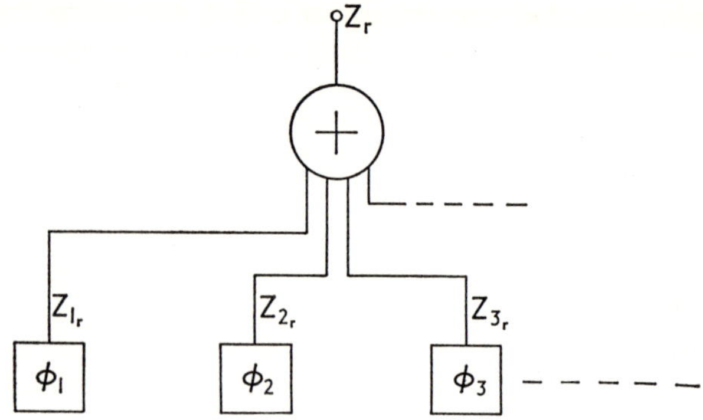

Fig. 4.1. *Realization of a product of different prime polynomials*

Polynomial Consisting of a Power of a Prime Polynomial

The combination of elementary feedback registers satisfying the equation $[\phi(D)]^K Z_r = 0$, where $\phi(D)$ is prime, is shown in Fig. 4.2. $R_1, R_2, \ldots, R_K$ are registers each separately satisfying the equation $\phi(D)Z_r = 0$, thus with $Z_{2r} = 0$, $\phi(D)Z_{1r} = 0$, etc. . . . Let $\phi(D) = 1 \oplus \phi_1(D)$ where 1 is D^0, then with $Z_{2r} = 0$, $Z_{1r} = \phi_1(D)Z_{1r}$. With $Z_{2r} \neq 0$, $Z_{1r} = \phi_1(D)Z_{1r} \oplus Z_{2r}$, so $Z_{2r} = [1 \oplus \phi_1(D)]Z_{1r} = \phi(D)Z_{1r}$. Similarly $Z_{3r} = \phi(D)Z_{2r}$, $Z_{4r} = \phi(D)Z_{3r}, \ldots, 0 = \phi(D)Z_{Kr}$.

Therefore $0 = [\phi(D)]^K Z_{1r}$, as required.

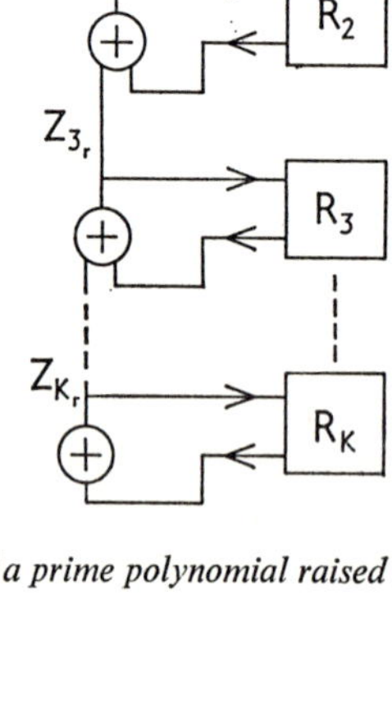

Fig. 4.2. *Realization of a prime polynomial raised to a power*

The sequences produced by the circuit of Fig. 4.2 can be classified as follows:

1. The all-zero sequence, when the initial conditions of $R_1 \dots R_K$ are zero.
2. Sequences satisfying $\phi(D)Z_r = 0$, when the initial conditions of R_1 are not all zero but those of $R_2 \dots R_K$ are.
3. Sequences satisfying $[\phi(D)]^2Z_r = 0$ but not $\phi(D)Z_r = 0$, when the initial conditions of R_2 (though not necessarily those of R_1) are not all zero, but those of $R_3 \dots R_K$ are.
4. Sequences satisfying $[\phi(D)]^3Z_r = 0$ but not $[\phi(D)]^2Z_r = 0$, when the initial conditions of R_3 (though not necessarily those of R_1 and R_2) are not all zero, but those of $R_4 \dots R_K$ are.

\- - - - - - - - - - - - -

$K + 1$. Sequences satisfying $[\phi(D)]^KZ_r = 0$ but not $[\phi(D)]^{K-1}Z_r = 0$, when the initial conditions of R_K (though not necessarily of $R_1 \dots R_{K-1}$) are not all zero, i.e. when all the registers R_1 to R_K are operative.

In effect, if registers R_1 to R_h are operative while R_{h+1} to R_K are inoperative we may regard R_h as producing a sequence which drives a linear binary filter comprising registers R_1 to R_{h-1}.

The periods of the different groups of sequences may be found in the manner explained at the beginning of this chapter. Thus sequence (1) has period one. Sequences (2) have period s, where s is the exponent of $\phi(D)$, i.e. the least value of t for which $\phi(D)$ divides $(0, t)$ exactly. Sequences (3) have period $2s$ since the binomial $(0, s)$ has a factor $\phi(D)$ and the next binomial containing $[\phi(D)]^2$ as a factor is $(0, 2s)$.

Sequences (4) have period $4s$ (not $3s$), because the next binomial containing $[\phi(D)]^3$ as a factor is $(0, 4s)$. This binomial in fact contains $[\phi(D)]^4$ as a factor, and so sequences (5) also have period $4s$. In the same way sequences (6), (7), (8), and (9) all have period $8s$, and sequences (10) to (17) have period $16s$, etc.

To enumerate the sequences in the various groups we use the example $\phi(D) = (0, 1, 2, 3, 4)$. This is a non-maximal prime polynomial, and the corresponding equation produces the all-zero sequence and three sequences of period five.

Sequences satisfying the equation $(0, 1, 2, 3, 4)^2$ but not $(0, 1, 2, 3, 4)$ have period ten. The former equation corresponds to an 8-stage register having $2^8 = 256$ states. Since 16 states are accounted for by sequences satisfying $(0, 1, 2, 3, 4)$, the remaining 240 states form 24 cycles each of length 10, giving 24 sequences of period 10.

Sequences satisfying $(0, 1, 2, 3, 4)^3$ but not $(0, 1, 2, 3, 4)^2$ have period 20.

The former equation corresponds to a register having $2^{12} = 4096$ states. Of these, 256 are accounted for above, so the remaining 3840 states form 192 cycles each of length 20. The cycle structure for the equation $(0, 1, 2, 3, 4)^3$ may therefore be written 1(1), 3(5), 24(10), 192(20).

The argument may be continued for higher powers.

The General Polynomial

The arrangement satisfying the general linear equation

$$[\phi_1(D)]^{a1}[\phi_2(D)]^{a2}[\phi_3(D)]^{a3} \dots Z_r = 0,$$

where $\phi_1(D)$, $\phi_2(D)$, $\phi_3(D)$, etc. are prime, is shown in Fig. 4.3. The

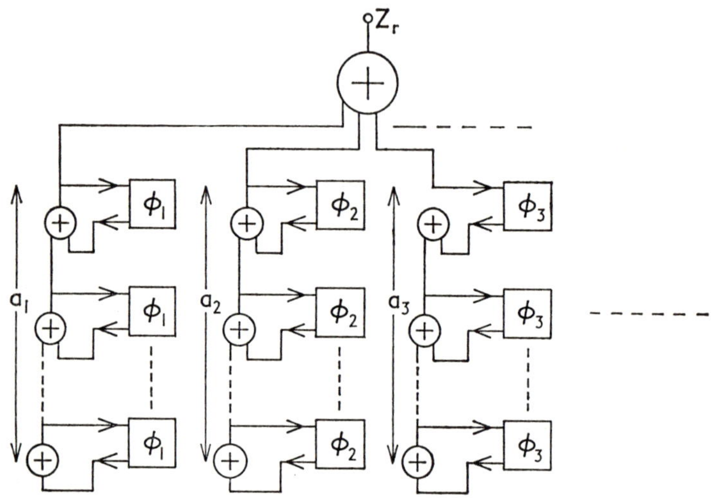

Fig. 4.3. *Realization of a general polynomial*

length and number of possible sequences in any given example can be worked out by a systematic application of the principles described above, but this becomes tedious as soon as the number of factors or the powers become at all large.

Chapter five

Linear Prime Polynomials

In this chapter the factorization of the binomial (0, s) into linear prime polynomials is considered. The equation (0, s) defines all sequences of period equal to or dividing s. Only a proportion of these, however, can be generated by a (non-cycling) linear feedback shift register. For example (0, 7) defines 20 sequences (of period one or seven), but only three of these can be generated by linear feedback registers.

Before proceeding, certain relevant theorems will be stated: some of these have already been noted.

1. A polynomial consisting of an even number of terms contains the factor (0, 1), since if (0, 1) is divided into it successive pairs of terms of the polynomial are divided out exactly. In particular (0, 1) is a factor of (0, s) for all s.
2. The periods of all non-zero sequences associated with a prime polynomial are the same, odd, and equal to the polynomial's exponent.
3. If $2^n - 1$ is prime, so also is n, for if $n = pq$, $2^n - 1 = 2^{pq} - 1 = (2p)^q - 1$, which is divisible by $2p - 1$. The converse is not true: if $2^n - 1$ is not prime, n may or may not be prime. If $2^n - 1$ is prime therefore, since n is also prime, n divides $2^n - 2$ by Fermat's theorem.
4. Direct division of $2^m - 1$ into $2^n - 1$ ($n > m$) shows that if m divides n, $2^m - 1$ divides $2^n - 1$, and conversely if $2^m - 1$ divides $2^n - 1$, m divides n. Thus if m does not divide n, $2^m - 1$ does not divide $2^n - 1$, and conversely.
5. Among the sequences associated with a prime polynomial raised to any power other than unity some have even period (see chapter four). The exponent of any polynomial containing a repeated prime factor is therefore even. It follows that the binomial (0, s) where s is odd cannot contain repeated prime factors.
6. A binomial (0, $2s$) factorizes into polynomials whose terms are twice the terms of the factors of (0, s). For example (0, 7) = (0, 1) (0, 1, 3) (0, 2, 3); (0, 14) = $(0, 7)^2$ = $(0, 1)^2$ $(0, 1, 3)^2$ $(0, 2, 3)^2$ = (0, 2) (0, 2, 6) (0, 4, 6); (0, 28) = $(0, 14)^2$ = $(0, 7)^4$ = (0, 4) (0, 4, 12) (0, 8, 12), and so on.
7. A prime polynomial having an exponent which equals or divides s must itself divide (0, s), and must therefore be one of the prime factors of (0, s). For s odd (0, s) contains as unrepeated factors all the prime polynomials whose exponents equal or divide s, since all these divide (0, s). Furthermore (0, s) cannot contain any other factor, for if such a factor

existed its exponent would equal or divide s as well. Thus $(0, s)$ *factorizes into* all the prime polynomials whose exponents equal or divide s. For example (0, 21) = (0, 1) (0, 1, 2) (0, 1, 3) (0, 2, 3) (0, 1, 2, 4,6) (0, 2, 4, 5, 6). The exponents of the factors taken in order are 1, 3, 7, 7, 21, and 21, these being the divisors of 21.

8. A celebrated extension of Fermat's theorem known as the Fermat–Euler theorem states that if a and m are relatively prime, $a^{\phi(m)} \equiv 1 \pmod{m}$, i.e. m divides $a^{\phi(m)} - 1$. For example, if $a = 3$ and $m = 10$, $\phi(m) = 4$ and 10 divides $3^4 - 1$. In this example a is a primitive root of m, i.e. m does not divide $a^K - 1$ where K is any number less than $\phi(m)$. However, if $a = 2$ and $m = 15$, $\phi(m) = 8$ and 15 divides not only $2^8 - 1$ but also $2^4 - 1$. Two is not therefore a primitive root of 15.

Suppose d is the least number for which $a^d \equiv 1 \pmod{m}$. Then $a^d \equiv a^{2d} \equiv a^{3d} \equiv \ldots \equiv 1 \pmod{m}$, and $a^{\phi(m)}$ must be one of these terms. Consequently $\phi(m)$ is a multiple of d. In particular, if $a = 2$ and $m = 2^n - 1$ the value $x = n$ satisfies the congruence $2^x \equiv 1 \pmod{2^n - 1}$ and is the least value to do so. Therefore $\phi(2^n - 1)$ is divisible by n.

Notice that in general n also divides the Euler function of some of the divisors of $2^n - 1$. For example, when $n = 8$, $2^n - 1 = 255$, whose divisors are 1, 3, 5, 15, 17, 51, 85, and 255. We see that 8 divides $\phi(255)$, $\phi(85)$, $\phi(51)$, $\phi(17)$, and $\phi(15)$, but not $\phi(5)$, $\phi(3)$, or $\phi(1)$. So long as n is the least value of x which satisfies the congruence $2^x \equiv 1 \left(\bmod \dfrac{2^n - 1}{K}\right)$, n divides $\phi\left(\dfrac{2^n - 1}{K}\right)$, but the above example in which 8 divides $\phi(15)$ shows that cases occur where n divides $\phi\left(\dfrac{2^n - 1}{K}\right)$ even though n is not the least value of x satisfying the above congruence. (In fact $2^4 \equiv 1 \pmod{15}$, and 8 divides $\phi(15)$ because $\phi\left(\dfrac{15}{4}\right)$ happens to be an even number.)

It is obvious from the Fermat–Euler theorem that every odd number divides a number of the form $2^n - 1$.

Number of Prime Polynomials of Degree n where $2^n - 1$ is Prime

In this case the binomial $(0, 2^n - 1)$ factorizes into (0, 1) and all the prime polynomials of degrees n, since no other polynomials having exponents dividing $2^n - 1$ can exist. There are therefore $\left(\dfrac{2^n - 2}{n}\right)$ of such polynomials, all of which are maximal. For example when $n = 5$, $2^n - 1 = 31$ which is prime, and there are $\dfrac{32 - 2}{5} = 6$ maximal polynomials of degree 5.

Number of Prime Polynomials of Degree n—the General Case

When $2^n - 1$ is not prime some polynomials of degree n will be maximal with exponent $2^n - 1$, and others will be non-maximal, with exponents dividing $2^n - 1$. The binomial $(0, 2^n - 1)$ will contain all these as factors, as well as others (maximal and non-maximal) of degrees of less than n and dividing n, since the exponents of these equal or divide $2^d - 1$, which itself divides $2^n - 1$. A non-maximal polynomial of degree n, however, cannot have an exponent s which divides or equals $2^d - 1$ (d/n and $d < n$) since $(0, 2^d - 1)$ factorizes into polynomials of degrees equal to or dividing d—the argument applied to polynomials of degree n evidently also applies to polynomials of degree d. The following examples clarify this:

1. For $n = 10$, $2^n - 1 = 1023$. The divisors of 10 are 1, 2, 5, and 10, and those of 1023 are 1, 3, 11, 31, 33, 93, 341, and 1023. The binomial (0, 1023) factorizes into:

(0, 1) — maximal, degree 1, exponent 1 — Exponent divides $2^1 - 1$
(0, 1, 2) — maximal, degree 2, exponent 3 — Exponent divides $2^2 - 1$ but not $2^1 - 1$
Six factors (0, . . ., 5) — maximal, degree 5, exponent 31 — Exponent divides $2^5 - 1$ but not $2^1 - 1$

(0, . . ., 10) — non-maximal, degree 10, exponent 11
Two factors (0, . . ., 10) — non-maximal, degree 10, exponent 33
Six factors (0, . . ., 10) — non-maximal, degree 10, exponent 93
Thirty factors (0, . . ., 10) — non-maximal, degree 10, exponent 341
Sixty factors (0, . . ., 10) — maximal, degree 10, exponent 1023

(these five:) Exponents divide $2^{10} - 1$ but not $2^5 - 1$, $2^2 - 1$ or $2^1 - 1$

2. For $n = 8$, $2^n - 1 = 255$. The divisors of 8 are 1, 2, 4, and 8, and those of 255 are 1, 3, 5, 15, 17, 51, 85, and 255. The binomial (0, 255) factorizes into:

(0, 1) — maximal, degree 1, exponent 1 — Exponent divides $2^1 - 1$
(0, 1, 2) — maximal, degree 2, exponent 3 — Exponent divides $2^2 - 1$ but not $2^1 - 1$

(0, 1, 2, 3, 4) — non-maximal, degree 4, exponent 5 (0, 1, 4)(0, 3, 4) — maximal, degree 4, exponent 15	Exponents divide $2^4 - 1$ but not $2^2 - 1$ or $2^1 - 1$

Two factors (0, . . ., 8) — non-maximal, degree 8, exponent 17 Four factors (0, . . ., 8) — non-maximal, degree 8, exponent 51 Eight factors (0, . . ., 8) — non-maximal, degree 8, exponent 85 Sixteen factors (0, . . ., 8) — maximal, degree 8, exponent 255	Exponents divide $2^8 - 1$ but not $2^4 - 1$, $2^2 - 1$ or $2^1 - 1$

We will use the first example to show how prime polynomials of given exponent may be enumerated. Consider the factorizations of (0, 1), (0, 3), (0, 11), (0, 31), (0, 33), (0, 93), (0, 341), and (0, 1023) in turn, remembering that each factorizes into all polynomials whose exponents divide 1, 3, . . ., 1023, respectively. Let $N(s)$ be the number of prime polynomials of exponent s. Equating degrees in each case we have:

$$\begin{aligned}
1 &= 1 \,.\, N(1)\\
3 &= 1 \,.\, N(1) + 2 \,.\, N(3)\\
11 &= 1 \,.\, N(1) + 10 \,.\, N(11)\\
31 &= 1 \,.\, N(1) + 5 \,.\, N(31)\\
33 &= 1 \,.\, N(1) + 2 \,.\, N(3) + 10 \,.\, N(11) + 10 \,.\, N(33)\\
93 &= 1 \,.\, N(1) + 2 \,.\, N(3) + 5 \,.\, N(31) + 10 \,.\, N(93)\\
341 &= 1 \,.\, N(1) + 10 \,.\, N(11) + 5 \,.\, N(31) + 10 \,.\, N(341)\\
1023 &= 1 \,.\, N(1) + 2 \,.\, N(3) + 10 \,.\, N(11) + 5 \,.\, N(31) + 10 \,.\, N(33)\\
&\quad + 10 \,.\, N(93) + 10 \,.\, N(341) + 10 \,.\, N(1023)
\end{aligned}$$

Defining $p(s)$ as the product of $N(s)$ and the degree of the polynomials of exponent s, the above equations became:

$$\begin{aligned}
1 &= p(1)\\
3 &= p(1) + p(3)\\
11 &= p(1) + p(3)\\
31 &= p(1) + p(31)\\
33 &= p(1) + p(3) + p(11) + p(33)\\
93 &= p(1) + p(3) + p(31) + p(93)\\
341 &= p(1) + p(11) + p(31) + p(341)\\
1023 &= p(1) + p(3) + p(11) + p(31) + p(33) + p(93) + p(341) + p(1023),
\end{aligned}$$

and we recognize the function $p(s)$ as the Euler function $\phi(s)$. In general, therefore, the number of prime polynomials of exponent s dividing $2^n - 1$ but not $2^K - 1 (K < n) = \frac{\phi(s)}{n}$, since all these are of degree n. In particular, the number of maximal polynomials of degree n is $\frac{\phi(2^n - 1)}{n}$.

In terms of degrees, the number $I(n)$ of prime ('irreducible') polynomials of degree n is given by $2^n - 1 = \sum_{d|n} dI(d)$, since $(0, 2^n - 1)$ factorizes into all the prime polynomials whose degrees divide n.

$$\text{Therefore } nI(n) = \sum_{d|n} \mu\left(\frac{n}{d}\right)(2^d - 1)$$

$$= \sum_{d|n} \mu\left(\frac{n}{d}\right)2^d - \sum_{d|n} \mu\left(\frac{n}{d}\right)$$

$$= \sum_{d|n} \mu\left(\frac{n}{d}\right)2^d, \text{ since the right-hand term is zero.}$$

$$\text{Therefore} \qquad I(n) = \frac{1}{n}\sum_{d|n} \mu\left(\frac{n}{d}\right)2^d = \frac{1}{n}\sum_{d|n} \mu(d)2^{n/d}.$$

$$\text{It follows that} \qquad \sum_{\substack{s/2^n-1 \\ s\nmid 2^K-1 (K<n)}} \phi(s) = \sum_{d|n} \mu(d)2^{n/d}.$$

The above formulae cover all possible odd exponents (and all possible degrees, therefore) since any odd number divides a number of the form $2^n - 1$, and prime polynomials can be found to generate (linear) sequences of any odd period.

Notice that $I(n)$ is the same as $\nu(n)$, the number of sequences generated by an n-stage cycling register.

Cyclotomic Polynomials

In the field of complex numbers the roots of $x^s - 1 = 0$ are $e^{2J(K/s)}$, $(K = 1, 2 \ldots, s)$. The complex ones among these occur in conjugate pairs. The roots for which K is prime to s are called primitive sth roots of unity, and there are $\phi(s)$ of these.

Consider $s = 15$. The divisors of 15 are 1, 3, 5, and 15, so $x^{15} - 1$ has binomial factors $x^d - 1$, where d may be any one of these divisors. The roots $a_K (K = 1, 2 \ldots, 15)$ of $x^{15} - 1 = 0$ may be arranged into groups satisfying $x^d - 1 = 0$ but not $x^{d'} - 1 = 0 (d' < d)$. Thus $a_{15} = e^{2\pi J(15/15)} = 1$ satisfies $x - 1 = 0$. The conjugate pair a_5, a_{10} satisfies $x^3 - 1 = 0$, but not $x - 1 = 0$. The conjugate pairs a_3, a_{12} and a_6, a_9 satisfy $x^5 - 1 = 0$, but not $x^3 - 1 = 0$ or $x - 1 = 0$, while the remaining conjugate pairs (a_1, a_{14}), (a_2, a_{13}), (a_4, a_{11}), and (a_7, a_8) satisfy $x^{15} - 1 = 0$ but not $x^5 - 1 = 0$, $x^3 - 1 = 0$ or $x - 1 = 0$.

The polynomials $P(1) = x - a_{15}$, $P(3) = (x - a_5)(x - a_{10})$, $P(5) = (x - a_3)(x - a_{12})(x - a_6)(x - a_9)$ and $p_{(15)} = (x - a_1)(x - a_{14})(x - a_2)(x - a_{13})(x - a_4)(x - a_{11})(x - a_7)(x - a_8)$, of degrees $\phi(1)$, $\phi(3)$, $\phi(5)$, and $\phi(15)$ respectively, are called cyclotomic polynomials of order 1, 3, 5, and 15 respectively, and we note that the coefficients in these polynomials are all real, and that the product $P(1) . P(3) . P(5) . P(15) . = x^{15} - 1$. In general the cyclotomatic polynomial $P(d)$ of order d has real coefficients and is of degree $\phi(d)$, and the roots of $P(d) = 0$ are the primitive roots of $x^d - 1 = 0$, i.e. the roots which satisfy $x^d - 1 = 0$ but not $x^{d'} - 1 = 0$ $(d' < d)$. Thus $x^s - 1$ may be expressed as the product $\prod_{d/s} P(d)$, from which it follows that $P(s) = \prod_{d/s} (x^d - 1)^{\mu(s/d)}$ (see chapter three).

It is clear that the cyclotomic polynomials $P(s)$, $(s/2^n - 1)$, are closely related to the factors (mod 2) of $(0, 2^n - 1)$. In the above example in which $n = 4$, for instance, $P(1) = x - 1$, $P(3) = x^2 + x + 1$, $P(5) = x^4 + x^3 + x^2 + x + 1$, and $P(15) = x^8 - x^7 + x^5 - x^4 + x^3 - x + 1$. When x is restricted to the values 0, 1 and polynomials are expressed module 2, these become respectively $x \oplus 1$, $x^2 \oplus x \oplus 1$, $x^4 \oplus x^3 \oplus x^2 \oplus x \oplus 1$, and $x^8 \oplus x^7 \oplus x^5 \oplus x^4 \oplus x^3 \oplus x \oplus 1$, i.e. (0, 1), (0, 1, 2), (0, 1, 2, 3, 4), and (0, 1, 3, 4, 5, 7, 8), using the previous notation, and we see that these last are, or factorize (module 2) into, the prime factors of $(0, 2^n - 1)$. In particular (0, 1, 3, 4, 5, 7, 8) factorizes into (0, 1, 4)(0, 3, 4), which are the two maximal polynomials of exponent 15. In general the cyclotomic polynomial (modulo 2) of order s factorizes (module 2) into all the prime polynomials of exponent s. (It follows directly that the number of maximal polynomials of degree n is $\phi(2^n - 1)/n$.)

While the above formula for $P(s)$ yields the product of all the prime polynomials of exponent s it does not give the prime polynomials themselves. The derivation of prime polynomials will be discussed later.

Chapter six

The Shift-and-Add Property and Remote-term Calculation

An important property of all *m*-sequences, known as the shift-and-add property, is that termwise addition of phase-shifted versions of a given *m*-sequence produces another phase-shifted version of the same sequence. This follows since if two sequences $\{x_r\}$ and $\{y_r\}$ obey the same equation $\phi(D)x_r = \phi(D)y_r = 0$, a sequence $\{z_r\}$ whose term $z_r = x_r \oplus y_r$ also obeys this equation. If $\phi(D)$ is maximal, all three are one and the same sequence. In general, therefore, if $\{y_r\}$ is an *m*-sequence, an integer *b* may be found for all integral values of *a* (other than zero or multiples of the period) such that $y_{r-b} = y_r \oplus y_{r-a}$. There is no direct functional relation between *a* and *b*. However, given *a*, *b* may be calculated in a number of ways, perhaps the easiest of which is shown in the following example: Suppose y_r is defined by the equation (0, 2, 5), and we wish to find *b*, given $a = 9$. We therefore want the single term $D^b y_r$ which equals $(1 \oplus D^9)y_r$, and so we divide (0, 2, 5) into (0, 9) as follows until a remainder consisting of a single term is reached:

```
          02456811
025 |  09
       025
       259
       247
       4579
       469
       567
       5710
       610
       6811
       81011
       81013
        1113
        111316
          16
```

and we see that $y_{r-16} = y_r \oplus y_{r-9}$

It follows that the same sequence results however many phase-shifted versions of a given m-sequence are added.

While no functional relation exists between a and b the case of $a = 1$ has been considered by Godfrey *et al.*[2] If $\{u_r\}$ is any binary sequence, the sequence $\{v_r\}$ given by $v_r = u_r \oplus u_{r-1}$ has a '1' at the start of each run of 0s or 1s in $\{u_r\}$. This follows since $v_r = 1$ when $u_r = \bar{u}_{r-1}$, and so $v_r = 1$ signals a change in the value of u_r. In the case of an m-sequence $\{y_r\}$ generated by an n-stage maximal register, the run of $(n - 1)$ 0s in the sequence formed by the termwise addition of y_r and y_{r-1} begins at the second '1' in the run of n 1s in $\{y_r\}$. The phase difference b between the sequences $\{y_r\}$ and $\{y_r \oplus y_{r-1}\}$ is therefore the number of epochs between the first of the $(n - 1)$ 0s and the second of the n 1s in $\{y_r\}$. The reader may easily verify this using for example the m-sequence satisfying (0, 2, 3, 4, 5), for which $b = 20$.

As is shown in the next section, having found the pair $(1, b)$, a set of pairs $(2, 2b)$, $(4, 4b)$, $(8, 8b)$, . . . (mod $2^n - 1$) follows. Other pairs cannot, however, be obtained in this way.

Remote-term Calculation

If a sequence $\{y_r\}$ is generated by an n-stage feedback shift register, remote-term calculation implies calculation of some term y_{r-b} in terms of $y_r, y_{r-1}, \ldots, y_{r-n}$, i.e. in terms of the outputs available from the register itself. The function of $y_r, \ldots, y_{r-n}$ so obtained obviously defines the logic needed to produce a replica of $\{y_r\}$ phase-shifted by b epochs. In the case of sequences generated by linear feedback shift registers, i.e. sequences which satisfy linear recurrence relations, remote-term calculation is quite simple, and two methods will be described by means of examples.

1. Since division of a polynomial $f_1(D)$ by another polynomial $f_2(D)$ which is zero amounts to successive subtraction of zero from $f_1(D)$, the remainder polynomial at every stage of division equals $f_1(D)$. Thus to obtain y_{r-15} in terms of $y_{r-5}, \ldots, y_r$, where y_r satisfies the equation $y_{r-5} \oplus y_{r-2} \oplus y_r = 0$, we divide (5, 2, 0) into (15) until a remainder polynomial of degree 5 or less is reached:

```
5, 2, 0 | 15
          151210
          ------
          1210
          1297
          ----
          1097
          1075
          ----
           95
           964
           ---
           654
           631
           ---
           5431
           520
           -----
           43210
```

and so y_{r-15} equals either $y_{r-5} \oplus y_{r-4} \oplus y_{r-3} \oplus y_{r-1}$, which are outputs from stages of the generating register, or $y_{r-4} \oplus y_{r-3} \oplus y_{r-2} \oplus y_{r-1} \oplus y_r$, which includes the feedback term y_r. In fact y_{r-15} equals every one of the remainders in the division process.

2. If a sequence satisfies $\phi(D)y_r = 0$ it also satisfies $\phi(D^2)y_r = 0$, $\phi(D^4)y_r = 0$, $\phi(D^8)y_r = 0$, etc., since in modulo 2, $\phi(D^2) = [\phi(D)]^2$, $\phi(D^4) = [\phi(D^2)]^2 = [\phi(D)]^4$, etc. If therefore $\phi(D)$ is zero, so are $\phi(D^2)$, $\phi(D^4)$, $\phi(D^4)$, $\phi(D^8)$, etc. For example, a sequence satisfying equation (0, 1, 6) also satisfies equations (0, 2, 12), (0, 4, 24), etc.

Suppose y_{r-56} is required in terms of $y_r, \ldots, y_{r-5}$, where y_r is generated by equation (0, 1, 6). The equations satisfied by y_r are:

(0, 1, 6) (given), (1, 2, 7), (2, 3, 8), etc.
(0, 2, 12), (1, 3, 13), (2, 4, 14), etc.
(0, 4, 24), (1, 5, 25), (2, 6, 26), etc.
(0, 8, 48), (1, 9, 49), (2, 10, 50), etc.
(0, 16, 96), (1, 17, 97), (2, 18, 98), etc.
etc.

Thus $56 = 48 + 8 = [(0 + 8), (8 + 8)] = (8, 16)$, since $48 = (0, 8)$.
$16 = 12 + 4 = [(0 + 4), (2 + 4)] = (4, 6)$, since $12 = (0, 2)$.
$8 = 6 + 2 = [(0 + 2), (1 + 2)] = (2, 3)$, since $6 = (0, 1)$.

Therefore $y_{r-56} = y_{r-8} \oplus y_{r-16} = (y_{r-2} \oplus y_{r-3}) \oplus (y_{r-4} \oplus y_{r-6}) = y_r \oplus y_{r-1} \oplus y_{r-2} \oplus y_{r-3} \oplus y_{r-4}$, since $6 = (0, 1)$.

An interesting method of economizing in modulo-two adders, both for producing phase-shifted versions of a given *m*-sequence and for producing *m*-sequences themselves, is described by Scholefield.[3] The principle of the method is to divide the *n* stages of the shift register into sub-registers containing suitable numbers *p*, *q*, *r*, etc. of stages, the input to any one sub-register other than the first being an appropriate linear combination of the outputs of the stages of the previous sub-register. The input to the first sub-register is the output from the last modulo-two adder. The method is particularly effective in the case of very long *m*-sequences in which a large number of adders would be required to produce a given phase-shift by the conventional method.

Non-maximal and Product Sequences

Both the above methods of remote-term calculation apply also to both non-maximal and non-prime polynomials, but the shift-and-add property applies universally only to maximal polynomials. Only for these do division sums of the form shown in the first section always lead to a single-term remainder, irrespective of the value of *a* in the binomial (0, *a*), or what is equivalent, if a polynomial of degree *n* and exponent *s* is divided into the monomial (0), binomials (*a*, *b*) containing between them all the numbers 1 to $(s - 1)$ occur as remainders only if $s = 2^n - 1$, i.e. only if the polynomial is maximal. If $s < 2^n - 1$ some of the numbers are missing, and the corresponding phase-shifted sequences are not formed by adding two relatively phase-shifted versions of the same sequence. Instead one of the other sequences in the cycle set results. The following examples show this:

1. Equation (0, 2, 3) generates the *m*-sequence 1110010, repeats. Numbers *a* and *b* satisfying $y_{r-b} = y_r \oplus y_{r-a}$ are (2, 3), corresponding to the generating equation itself, (4, 6) and (1, 5). All numbers 1 to 6 are present.
2. Equation (0, 1, 2, 3, 4) generates non-maximal prime sequences $\{u_r\} =$ 11110, repeats, $\{v_r\} =$ 10100, repeats, and $\{w_r\} =$ 10001, repeats. No numbers *a* and *b* satisfy $y_{r-b} = y_r \oplus y_{r-a}$, and $\{u_r \oplus u_{r-1}\} \rightarrow \{w_r\}$, $\{u_r \oplus u_{r-2}\} \rightarrow \{v_r\}$, and so on.
3. Equation (0, 1, 5) = (0, 1, 2)(0, 2, 3) generates sequences $\{u_r\}$ of period 21, $\{v_r\}$ of period 7, and $\{w_r\}$ of period 3. Applying $y_{r-b} = y_r \oplus y_{r-a}$ to $\{u_r\}$, we find (*a*, *b*) can assume only the following values: (1, 5), corresponding to the generating equation, (2, 10), (4, 20), (8, 19), (16, 17), and (11, 13). Thus, for example $\{u_r \oplus u_{r-11}\} \rightarrow \{u_r\}$ since $u_r \oplus u_{r-11} = u_{r-13}$, but $\{u_r \oplus u_{r-7}\} \rightarrow \{w_r\}$, $\{u_r \oplus u_{r-3}\} \rightarrow \{v_r\}$, etc. Notice that all the numbers *a* and *b* above are prime to the exponent, 21.

More generally, the terms $y_r \ldots y_{r-n}$ available from an *n*-stage linear feedback register may be independently linearly combined in $2^n - 1$ ways (excluding the 'combination' of no terms). Since no two of these combinations yield the same output for all values of r, only if $\{y_r\}$ is maximal are there enough possible phase shifts to allow all the combinations to produce phase-shifted versions of the same sequence.

Mechanization of Polynomial Division

Since polynomial division forms the basis of most calculations relating to linear sequences it is worth describing one of a number of electronic polynomial dividers given, for example, by Peterson.[4]

Consider first the division of $D^4 \oplus D \oplus 1$ into D^0, D^1, D^2, etc., in turn. In detached exponent notation the quotients and remainders are respectively the polynomials:

quotient	–	–	–	–	0	1	2	3, 0	4, 1, 0	. . .
remainder	0	1	2	3	1, 0	2, 1	3, 2	3, 1, 0	2, 0	. . .

We see that if the highest number in the remainder is less than 3, the next remainder is formed by adding one to each number in the remainder, and the next quotient is formed similarly by adding one to each number in the quotient. If the highest number in the remainder is 3, however, the next remainder is formed by adding one to each number in the remainder as before, but also subtracting the dividing polynomial (4, 1, 0) from the result. The next quotient is formed by adding one to each number in the quotient as before, and adding the number 0 as well.

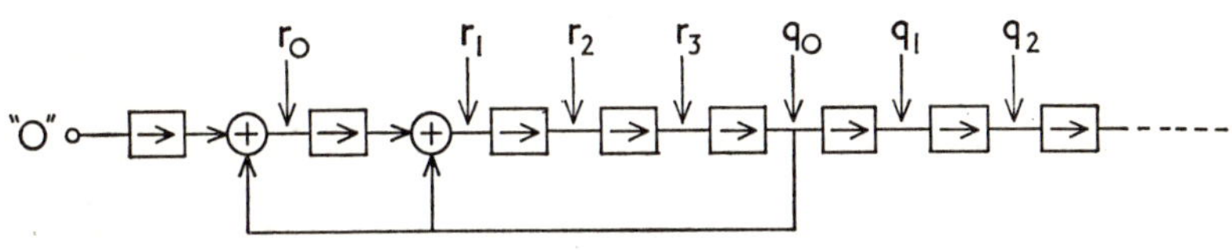

Fig. 6.1. *Circuit for calculating* $\dfrac{D^n}{D^4 \oplus D \oplus 1}$

The circuit of Fig. 6.1, in which all the stores are simultaneously clocked, realizes the above procedure. The initial output of the first store is '1' and that of all the others is '0'. After n clock pulses D^n has been divided by $D^4 \oplus D \oplus 1$ to yield the remainder $r_0 \oplus r_1 D \oplus r_2 D^2 \oplus r_3 D^3$ and the quotient $q_0 \oplus q_1 D \oplus \ldots \oplus q_{n-4} D^{n-4}$ (zero for $n < 4$), where the coefficients are the logic levels at the corresponding points in the circuit.

The general division sum $\frac{g(D)}{f(D)} = q(D) + \frac{r(D)}{f(D)}$, where

$$g(D) = g_0 \oplus g_1 D \oplus \ldots \oplus g_m D^m,$$
$$f(D) = f_0 \oplus f_1 D \oplus \ldots \oplus f_{n-1} D^{n-1} \oplus D^n,$$
$$q(D) = q_0 \oplus q_1 D \oplus \ldots \oplus q_{m-n} D^{m-n} \; (= 0 \text{ for } m < n)$$
$$r(D) = r_0 \oplus r_1 D \oplus \ldots \oplus r_{n-1} D^{n-1},$$

is completed by the circuit of Fig. 6.2. after m clock pulses. The initial outputs of the dividend register stores, reading from the left, are $g_0, g_1, \ldots, g_m$ and those of all the others are zero. The switches $s_0, s_1, \ldots, s_{n-1}$ are open or closed according as $f_0, f_1, \ldots, f_{n-1}$ are 0 or 1 respectively.

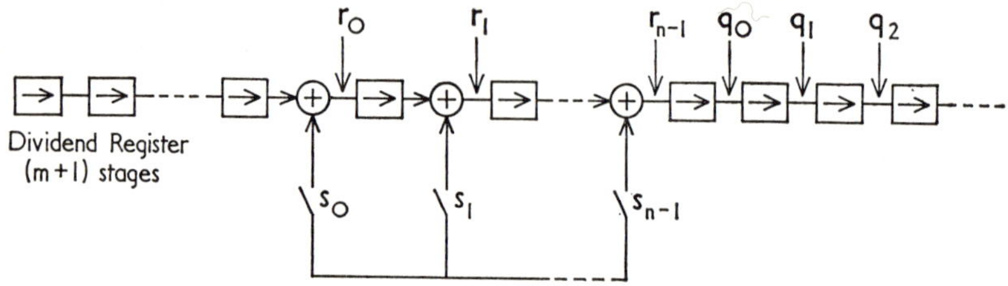

Fig. 6.2. *Circuit for calculating* $\frac{g(D)}{f(D)}$

It is instructive to work through the action for a simple case such as:

$$\frac{D^3 \oplus D^2 \oplus 1}{D^2 \oplus D \oplus 1} = D \oplus \frac{D \oplus 1}{D^2 \oplus D \oplus 1}$$

The above dividing circuit may be regarded as a binary sequence filter of the type described in chapter one. The quotient output from the circuit of Fig. 6.1, treated as a sequence, is the filter's impulse response. In the circuit of Fig. 6.2 the case of $f(D) = g(D)$ corresponds to using the filter's null sequence as its input. The reader should also recall the section in chapter four headed 'Significance of Quotient' in this connection.

Generation of Sequences of Prescribed Period

By virtue of the shift-and-add property of m-sequences, a sequence of any prescribed period v may be generated by a suitably modified n-stage maximal register, where $1 \leqslant v \leqslant 2^n - 1$, though obviously for $v \leqslant 2^{n-1} - 1$ an $(n - 1)$-stage register would suffice. Providing the intervening count does not have to be in binary, a shift register generating a sequence of period v may be used as a counter (modulo v). Since all the stages are clocked simultaneously there is no propagation delay, as in conventional serial counters, and such a counter therefore has the advantages of being very fast as well as simple.

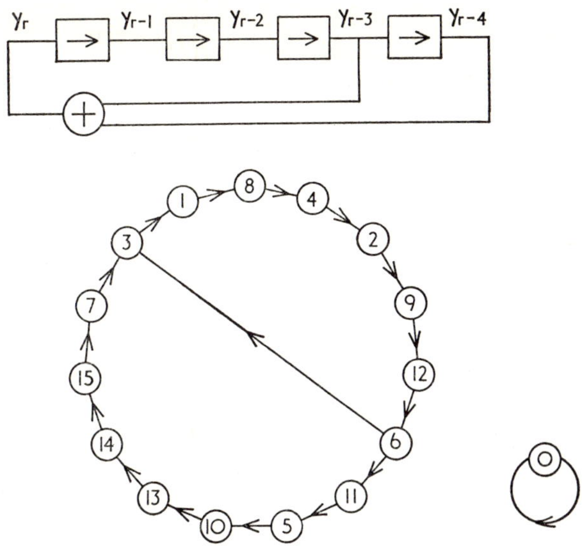

Fig. 6.3. *Register and state diagram for sequence satisfying* $y_r = y_{r-3} \oplus y_{r-4}$

Consider the *m*-sequence 000100110101111, repeats, generated by the register shown in Fig. 6.3. The state-transition diagram is also shown, the binary equivalent of the decimal number representing a state being the number $(y_{r-1}y_{r-2}y_{r-3}y_{r-4})$. Successive digits of the sequence are therefore fed into this number from the left.

Suppose the sequences $\{y_r\}$ and $\{y_{r+9}\}$ are added term by term to produce the sequence $\{y_{r+7}\}$, as shown in Fig. 6.4. The patterns above and below 0001 in $\{y_{r+9}\}$ differ only in their right-hand digit, so the register may be made to 'jump' from state 0110 in sequence $\{y_r\}$ to the state 1100 in the sequence $\{y_{r+7}\}$, instead of the next state 1101 in $\{y_r\}$, by causing the pattern 0110 to produce an additional '1' over and above the normal feedback digit, and feeding it as an extra input to the modulo-two adder in the register feedback circuit. In this way state 6 in Fig. 6.3 is followed by state 3, seven places beyond the state 11 which would normally have succeeded

$[y_r]$:	0	0	0	1	0	0	1	1	0	1	0	1	1	1	1
$[y_{r+9}]$:	1	0	1	1	1	1	0	0	0	1	0	0	1	1	0
$[y_{r+7}]$:	1	0	1	0	1	1	1	1	0	0	0	1	0	0	1

Fig. 6.4. $\{y_r\} \oplus \{y_{r+9}\} = \{y_{r+7}\}$

state 6, and the cycle length has been reduced from 15 to $15 - 7 = 8$. Fig. 6.5 shows the modified register and the corresponding state diagram.

The above process is known variously as sequence truncation or the jump technique. Since for every a in the equation $y_{r+a} = y_r \oplus y_{r+b}$ there is a corresponding b if $\{y_r\}$ is an m-sequence, any length from 1 to $2^n - 1$ can be produced in this way from an n-stage maximal register. For example, if $n = 5$ and we require a cycle of length 20, we need a word detector which causes a jump of $(2^5 - 1) - 20 = 11$ places. Using a generator satisfying

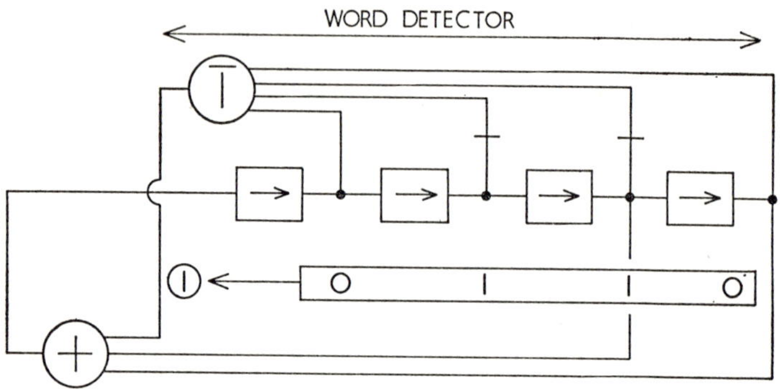

(a)

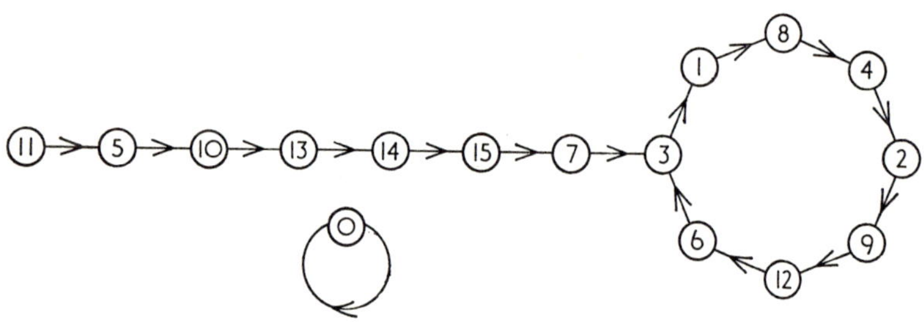

(b)

Fig. 6.5. *Modified register producing truncated sequence*

$y_r = y_{r-2} \oplus y_{r-5}$ we find $y_{r+11} = y_r \oplus y_{r+23}$. Writing $\{y_r\}$, $\{y_{r+23}\}$, and $\{y_{r+11}\}$ one above the other shows that the pattern 00001 in $\{y_{r+23}\}$ lies under the pattern 11101 in $\{y_r\}$. The 'word' to be detected is therefore the preceding pattern 01110, and so the required word detector logic is $\overline{y_{r-1} + \bar{y}_{r-2} + \bar{y}_{r-3} + \bar{y}_{r-4} + y_{r-5}}$.

It may be noted that if the detector produces a '1' from the $n - 1$ digits $y_{r-1}, y_{r-2}, \ldots y_{r-n+1}$, thereby leaving y_{r-n} as $\oplus y_{r-n}$ in the sequence equation, the major cycle in the state diagram splits into two separate cycles, one of the required length and the other consisting of the omitted

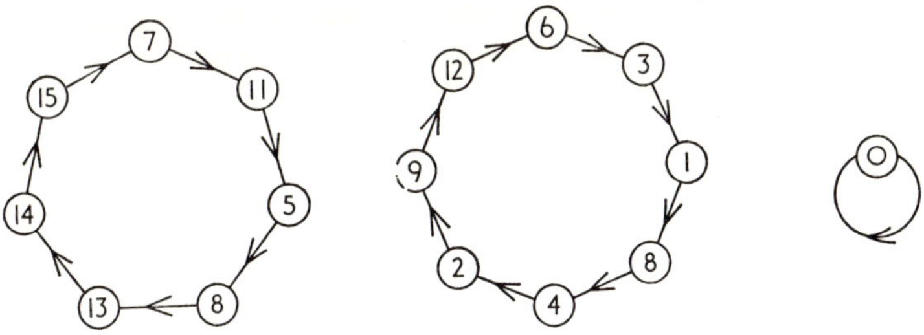

Fig. 6.6. m-*sequence split into two separate major sequences*

states. Using the example of Fig. 6.4, if the pattern $x110$ produces an extra '1', state 6 jumps to state 3 as before, but now also state 7 jumps to state 11, thereby leading to the state diagram shown in Fig. 6.6.

A maximal n-stage register may be modified as shown in Fig. 6.7 to produce a *full-length sequence* of period 2^n, i.e. a sequence containing in one period all the n-bit binary numbers 0 to $2^n - 1$. When the register reaches the state 00 . . . 01 the feedback digit is 0 instead of 1 by virtue of the 1 output from the NOR gate, and the next state is therefore 00 . . . 00. The feedback digit is then 1, so the next state after this is 10 . . .00. Thereafter, until the state 00 . . . 01 is again reached, the register operates in the linear mode. In this way the all-zero state is included in the sequence.

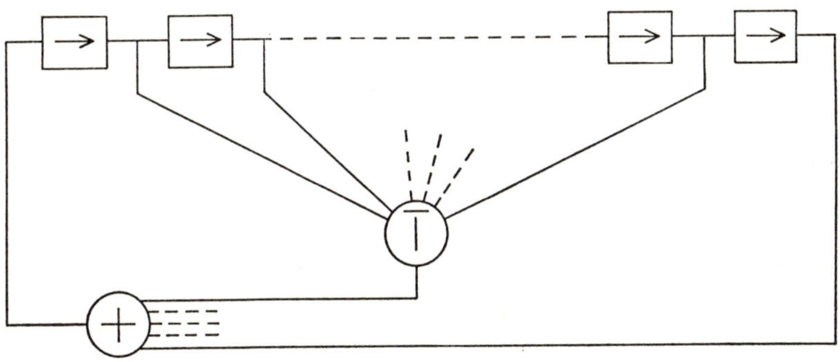

Fig. 6.7. *Full-length sequence generator*

Chapter seven

Sequence Sampling

Uniform sampling, or decimation, of a sequence $\{y_r\}$ implies forming a new sequence $\{z_r\}$ by selecting every mth term of $\{y_r\}$. In this chapter we consider some aspects of sampling, especially those associated with sequences satisfying linear recurrence relations. More general aspects are considered in chapter nine. The following are a few properties of sampled sequences:

1. If $\{y_r\}$ has period v, and m is prime to v ('proper decimation'), $\{z_r\}$ also has period v and is either a phase-shifted version of $\{y_r\}$ or a different sequence altogether. In the former case m is called a multiplier of $\{y_r\}$. If m is not prime to v, $\{z_r\}$ has period (LCM of m and v) $\div\ m$, and therefore necessarily differs from $\{y_r\}$. For example $\{y_r\} = 100110$, repeats, has period 6. The only proper decimation occurs for $m = 5$ and yields $\{z_r\} =$ 101100, repeats, which is a different sequence. On the other hand $\{y_r\} =$ 111010, repeats, also has period 6, and proper decimation yields $\{z_r\} =$ 101011, repeats, which is a phase-shifted version of $\{y_r\}$.

2. If m is prime to v, the numbers $0, m, 2m, \ldots, (v-1)m \pmod v$ are the numbers $0, 1, 2, \ldots, v-1$ in a different order, since it is not possible for two numbers gm and hm (g and $h < v$) both to be congruent (mod v) to the same number. Therefore the terms $z_0 = y_0$, $z_1 = y_m$, $z_2 = y_{2m}, \ldots$, $z_{v-1}, = y_{(v-1)m}$ of $\{z_r\}$ are merely a rearrangement of the terms $y_0, y_1, y_2, \ldots, y_{v-1}$ of $\{y_r\}$. In particular $\sum_{r=1}^{v} z_r = \sum_{r=1}^{v} y_r$, and the 'autocorrelation' $C_z(h)$ of $\{z_r\}$, defined as
$\frac{1}{v}\sum_{r=1}^{v} z_r z_{r+h}$, equals $\frac{1}{v}\sum_{r=1}^{v} y_{mr}y_{m(r+h)} = \frac{1}{v}\sum_{r'=1}^{v} y_{r'}y_{r'+h'}$, where $r' = mr$ (mod v) and $h' = mh \pmod v$, and therefore equals the autocorrelation $C_y(h')$ of $\{y_r\}$.

For example let $y_r = y_0y_1 \ldots y_6$, repeats with period 7. For $m = 3$, $z_r = y_0y_3y_6y_2y_5y_1y_4$, repeats. Thus $\sum_{r=1}^{7} y_r = \sum_{r=1}^{7} z_r$. For $h = 4$, $C_z(4) = \frac{1}{7}(y_0y_5 + y_3y_1 + y_6y_4 + y_2y_0 + y_5y_3 + y_1y_6 + y_4y_2)$. Now $h' = 3 \times 4 \pmod 7 = 5$, and $C_y(h') = C_y(5) = \frac{1}{7}(y_0y_5 + y_1y_6 + y_2y_0 + y_3y_1 + y_4y_2 + y_5y_3 + y_6y_4)$, which is the same as $C_z(4)$.

A case of special importance is when y_r has 'two-level autocorrelation', i.e. when

$$vC_y(h) = a \text{ for } h \equiv 0 \pmod v$$
$$= \beta \text{ for } h \not\equiv 0 \pmod v$$

where β is a constant independent of h. In this case $vC_z(h)$ also equals a for $h \equiv 0 \pmod v$ and β for $h \not\equiv 0 \pmod v$, and so $\{z_r\}$ has the same two-level autocorrelation as $\{y_r\}$. Sequences possessing two-level autocorrelation are called 'perfect sequences'.

As an example the sequence $\{y_r\} = 11-1111-1-1-11-1$, repeats with period 11, which has 3 as a multiplier, has two-level autocorrelation given by

$$11 \times C_y(h) = 11(h \equiv 0, \bmod 11)$$
$$= -1(h \not\equiv 0, \bmod 11)$$

The sequence $\{z_r\}$ obtained by sampling every other term of $\{y_r\}$ is $1-11-1-1-1111-11$, repeats, and has the same autocorrelation as $\{y_r\}$.

3. If m is a multiplier of $\{y_r\}$, since every term of $\{y_r\}$ appears somewhere in $\{z_r\}$, the same sequence $\{z_r\}$ results whatever term of $\{y_r\}$ sampling begins at.

4. If m is a multiplier of $\{y_r\}$ and sampling $\{y_r\}$ every mth term yields the sequence $\{z_r\}$, it is possible, subject to certain conditions, to find a term of $\{y_r\}$ to begin sampling at such that $\{z_r\}$ is termwise identical to $\{y_r\}$. This fact is cardinal in the construction of sequences possessing multipliers.

Consider a sequence $\{y_r\}$ of period 10 whose terms are $y_0, y_1, \ldots, y_9$. The only numbers prime to 10 are 3, 7, and 9. Suppose $m = 3$ is a multiplier. The sequence $\{z_r\}$ formed by sampling every third term of $\{y_r\}$, starting from y_0, is $y_0 y_3 \ldots y_4 y_7$, repeats. Suppose $\{z_r\}$ is the sequence $\{y_r\}$ phase-shifted (forward) by four, namely $y_4 y_5 \ldots y_2 y_3$, repeats. Thus $y_0 = y_4, y_3 = y_5$, etc., leading to the equations $y_0 = y_2 = y_4 = y_8 = a$, say, and $y_3 = y_5 = y_7 = y_9 = b$, say, with y_1 and y_6 as 'free' terms. $\{y_r\}$ is therefore the sequence $a\ y_1\ a\ b\ a\ b\ y_6\ b\ a\ b$, repeats. Since y_1 and y_6 are arbitrary they represent recognizable 'landmarks' in the sequence, and if sampling starts at either of these terms $\{z_r\}$, being the same sequence as $\{y_r\}$, is termed identical to $\{y_r\}$.

The phase shifts for which free terms exist are 0, 2, 4, 6, and 8, and evidently $\{y_r\}$ is of the same form whichever of these phase shifts is prescribed. We see therefore that if it is possible for $\{y_r\}$ and $\{z_r\}$ to be termwise identical, the form of $\{y_r\}$ (and $\{z_r\}$) is completely determined by its period and multiplier.

In general, if a sequence of period v has a multiplier m, and sampling from a given term, say y_0, yields the same sequence shifted forward γ places, free terms y_r occur only if γ is such that the congruence $(m - 1)r \equiv m\gamma \pmod{v}$ can be satisfied. For the above case this congruence becomes $2r \equiv 3 \times 4 \equiv 2 \pmod{10}$, and it is satisfied by $r = 1$ and 6, as found. Note that the congruence can be satisfied for all γ if v is prime or if $m = 2$.

Unless a solution to the above congruence exists there are no free terms, and so in general there is no point in $\{y_r\}$ from which to sample to yield a termwise identical sequence. With particular choice of variables, however, this may not be true. For example with $v = 10$, $m = 3$, and $\gamma = 1$, $\{y_r\}$ and $\{z_r\}$ have the general form $a\ a\ b\ b\ c\ b\ b\ a\ a\ c$, repeats, which does possess the termwise repeatability property.

The following examples demonstrate the construction of sequences having multipliers:

(*a*) Construct a sequence $\{y_r\}$ of period 7 having a multiplier 2. Since $m = 2$ there is a term from which to sample $\{y_r\}$ to yield a termwise identical $\{z_r\}$. Call this term y_0. Then $\{z_r\}$ becomes $y_0y_2y_4y_6y_1y_3y_5$, giving equations $y_1 = y_2 = y_4 = a$, say, and $y_3 = y_5 = y_6 = b$, say, yielding the sequence $y_0\ a\ a\ b\ a\ b\ b$, repeats. Suppose $a = 1$ and $b = 0$, for $y_0 = 1$ this becomes 1110100, repeats (an *m*-sequence), and for $y_0 = 0$ it becomes 0110100, repeats.
(*b*) Construct a sequence $\{y_r\}$ of period 11 having multiplier 3. Since v is prime there is a point from which to sample $\{y_r\}$ to yield a termwise identical $\{z_r\}$. Call this term y_0. Then z_r becomes $y_0y_3 \ldots y_8$, repeats, giving equations $y_1 = y_3 = y_4 = y_5 = y_9 = a$, say, and $y_2 = y_6 = y_7 = y_8 = y_{10} = b$, say, yielding the sequence $y_0\ a\ b\ a\ a\ a\ b\ b\ b\ a\ b$, repeats. With $y_0 = a = 1$ and $b = 0$, for example, it becomes 11011100010, repeats (a 'quadratic residue' sequence).

5. *Cosets*. If m is a multiplier of a sequence $\{y_r\}$ of period v, then so are $m^2 . m^3 \ldots \pmod{v}$, and the sequences $\{y_{mr}\}$, $\{y_{m^2r}\}$, etc., are all phase-shifted versions of $\{y_r\}$, where the notation $\{y_{qr}\}$ implies a sequence formed by sampling $\{y_r\}$ every qth term. The numbers $1, m, m^2, \ldots \pmod{v}$ form a 'coset', and the terms $y_1, y_m, y_{m^2}, \ldots$, of $\{y_r\}$ are all equal. In effect a coset may be defined as a set of equivalent sampling intervals. In particular if $\{y_{mr}\}$ is termwise identical to $\{y_r\}$, the sequences $\{y_r\}$, $\{y_{mr}\}$, $\{y_{m^2r}\}$, etc. are all termwise identical.

If l is not among $m, m^2, \ldots \pmod{v}$, $\{y_{lr}\}$ is a different sequence but the numbers $l, ml, m^2l, \ldots \pmod{v}$ form another coset because the sequences $\{y_{lr}\}$, $\{y_{(ml)r}\}$, $\{y_{(m^2l)r}\}$ are all phase-shifted versions of each other. (For example, $\{y_{(m^2l)r}\} = \{y_{l(m^2r)}\} = \{y_{lr}\}$.)

Thus the period and multiplier of a sequence define a coset structure.

For example with $v = 13$ and $m = 3$ the coset structure and corresponding sequences are:

1, 3, 9:	y_0 *a b a c b b d d a c d c*	
2, 6, 5:	y_0 *b c b d c c a a b d a d*	$(l = 2)$
4, 12, 10:	y_0 *c d c a d d b b c a b a*	$(l = 4)$
7, 8, 11:	y_0 *d a d b a a c c d b c d*	$(l = 7)$

where y_0 is the free term and *a*, *b*, *c*, and *d* are constants.

Notice that any of these sequences can be formed by suitably sampling any other. Note, also, that since the free term is the starting-point of all the sequences, $y_1 = y_3 = y_9$, $y_2 = y_6 = y_5$, $y_4 = y_{12} = y_{10}$, and $y_7 = y_8 = y_{11}$ in all the sequences.

Application of Sampling to Linear Sequences

As indicated in chapter six, a sequence which satisfies a linear equation $\phi(D)y_r = 0$ also satisfies $[\phi(D)]^2 y_r = \phi(D^2)y_r = 0$. Now $\phi(D^2)y_r = 0$ is equivalent to the equation $\phi(D)z_r = 0$, where $\{z_r\}$ is the sequence obtained by sampling $\{y_r\}$ every other term. Thus linear sequences of odd period all have the multiplier 2, and sequences formed by sampling a linear sequence $\{y_r\}$ of odd period every second, fourth, eighth, etc., term all satisfy the equation satisfied by $\{y_r\}$.

Consider the sequences produced by a five-stage maximal register. These all have period 31, and so their coset structure is

1, 2, 4, 8, 16
3, 6, 12, 24, 17
5, 10, 20, 9, 18
7, 14, 28, 25, 19
11, 22, 13, 26, 21
15, 30, 29, 27, 23

As shown in chapter five there are six *m*-sequences of period 31. Each of these corresponds to one of the cosets, since if any one is sampled every *i*th term ($i \not\equiv 0$, mod 31), another sequence of period 31 results which is either a phase-shifted version, or one of the other five *m*-sequences.

The six *m*-sequences can be constructed *ab initio* from the cosets along the lines indicated above (and described in detail by Golomb[5]), but if any of the *m*-sequences is known (e.g. from its characteristic polynomial) the other five can be derived by sampling it every third, fifth, seventh, eleventh, and fifteenth term. Having found the other sequences, their characteristic polynomials may be found. In other words, given one maximal polynomial of given degree, all the others of the same degree may be found via sequence sampling.

As an example, suppose $\{y_r\}$ is an *m*-sequence of period 31 satisfying the equation (0, 2, 5). Sampling this every third term gives a sequence $\{z_r\}$ whose terms are . . ., $y_{r-6}, y_{r-3}, y_r, y_{r+3}, \ldots$ We require an equation of fifth degree for $\{z_r\}$, so we have to calculate $y_{r-15}, y_{r-12}, y_{r-9}, y_{r-6}, y_{r-3}$, and y_r in terms of $y_{r-5}, y_{r-4}, y_{r-3}, y_{r-2}, y_{r-1}$, and y_r. This is accomplished by remote-term calculation, and we find:

$$z_{r-5} = y_{r-15} = y_{r-5} \oplus y_{r-4} \oplus y_{r-3} \oplus y_{r-1}$$
$$z_{r-4} = y_{r-12} = y_{r-3} \oplus y_{r-2} \oplus y_{r-1}$$
$$z_{r-3} = y_{r-9} = y_{r-4} \oplus y_{r-3} \oplus y_{r-1}$$
$$z_{r-2} = y_{r-6} = y_{r-3} \oplus y_{r-1}$$
$$z_{r-1} = y_{r-3}, \text{ and } z_r = y_r.$$

Solving for y_r, y_{r-2}, and y_{r-5} we have:

$$y_r = z_r, \quad y_{r-2} = z_{r-2} \oplus z_{r-4}, \quad \text{and} \quad y_{r-5} = z_{r-3} \oplus z_{r-5}.$$

So the equation (0, 2, 5) for $\{y_r\}$ leads to the equation (0, 2, 3, 4, 5) for $\{z_r\}$.

Complete Coset Structure

The non-zero sequences satisfying the binomial equation (0, 31) comprise six *m*-sequences of period 31 and one *m*-sequence of period one (corresponding to the factor (0, 1)). Adding a degenerate coset 31 to the coset structure given above accounts for the latter sequence, since sampling (from a '1') any one of the *m*-sequences at interval 31 yields the all-ones sequence.

Consider the binomial $(0, 2^n - 1)$ in which $n = 4$, and therefore $2^n - 1$ is not prime. The complete coset structure is:

15
1, 2, 4, 8
3, 6, 12, 9
5, 10
7, 14, 13, 11

Now (0, 15) factorizes into (0, 1) yielding the *m*-sequence of period 1, (0, 1, 2) yielding the *m*-sequence of period 3, (0, 1, 2, 3, 4) yielding three non-maximal prime sequences of period 5, (0, 1, 4) and (0, 3, 4) giving *m*-sequences of period 15. It readily follows that when $v = 2^n - 1$ the number of cosets associated with period $v = \sum_{d|n} I(d)$.

Sampling either of the *m*-sequences of period 15 every fifteenth term (from a '1') yields the m-sequence of period 1, sampling every fifth or tenth term yields the m-sequence of period 3. Sampling every third, sixth, twelfth, or ninth term yields the three non-maximal sequences of period 5, and sampling every seventh, fourteenth, thirteenth, or eleventh term yields the other m-sequence of period 15. Thus, by suitable choice of sampling period, all the sequences associated with all the prime factors of (0, 15) may be derived from one m-sequence of period 15. This is evidently true whatever the value of n.

Sampling of Product Sequences

If $\{y_r\}$ satisfies $\phi(D)y_r = 0$ where $\phi(D)$ is a product of unrepeated prime polynomials, we saw in chapter four that $\{y_r\}$ is formed by termwise addition of sequences $\{u_r\}$, $\{v_r\}$, $\{w_r\}$, etc., associated with each factor of $\phi(D)$. The sequence formed by sampling $\{y_r\}$ is therefore the termwise addition of sequences formed by simultaneously sampling $\{u_r\}$, $\{v_r\}$, $\{w_r\}$, etc. The polynomial associated with it is therefore the product of the polynomials associated with each sampled component sequence. For example, if $\phi(D)$ is (0, 1, 5) = (0, 1, 2)(0, 2, 3), $\{y_r\}$ has period 21, 7, or 3. Sampling the two component sequences every third term produces sequences satisfying equations (0, 1) and (0, 1, 3), and so the resulting sequence satisfies (0, 1)(0, 1, 3) = (0, 2, 3, 4). Sampling every fifth term yields a sequence satisfying (0, 1, 2)(0, 1, 3) = (0, 4, 5), etc.

Chapter eight

Noise-like Properties of *m*-Sequences

The major application of *m*-sequences depends on their so-called pseudo-random or pseudo-noise property. In so far as a feedback shift register sequence is completely determined by the generating equation and the state of the register at any given epoch it cannot be described as truly random. Furthermore, since a sequence's instantaneous amplitude is either 1 or 0 (or -1 or $+1$, depending on the choice of datum) it is certainly not noise-like in this respect. A noise waveform, unless its spectrum is very restricted indeed, has a gaussian amplitude distribution. Finally, the successive digits of a sequence are normally separated by multiples of a fixed clock period and exhibit a periodicity which is an exact multiple of the clock period. Consequently the spectrum of a sequence, in contrast to that of noise, is not continuous but consists of discrete frequency components. Nevertheless a long *m*-sequence does satisfy certain randomness criteria.

Criteria of Randomness

Suppose a coin is tossed a large number N of times, the outcome of any one spin being as likely to be heads as tails. The sequence of heads and tails so formed, though random, will exhibit the following characteristics:

(i) $$\operatorname*{Lt}_{N\to\infty}\left[\frac{\text{number of heads}}{\text{number of tails}}\right]=1,$$

(ii) $$\operatorname*{Lt}_{N\to\infty}\left[\frac{v(i)}{v(i+1)}\right]=2,$$

where for a given value of N, $v(i)$ is the number of runs of i heads, flanked at either end by a tail, in the sequence. This follows from the fact that if i heads have been thrown, the probability that another head followed by a tail will be thrown, to make a run of $i+1$ heads, is exactly half the probability that a tail will be thrown making a run of i heads.

(iii) The above two characteristics are not sufficient to describe the randomness of the sequence, since it is possible to construct a sequence satisfying both properties which is 'clearly' not random, e.g.:

HHH	TTT	HH	TT	HH	TT	H	T	H	T	H	T	H	T
3		2		2		1		1		1		1	

Suppose '1' is assigned to a head and '-1' to a tail. Any pair of distinct terms in the coin-tossing sequence may be $(1, 1)$, $(1, -1)$, $(-1, 1)$, or $(-1, -1)$ with equal likelihood. Consequently if the product of two terms y_r and y_{r+h} is formed, where h is a fixed integer, the sum

$$C(h) = \operatorname*{Lt}_{j\to\infty} \frac{1}{2j+1} \sum_{r=-j}^{+j} y_r y_{r+h} \begin{matrix} = 0 & (h \neq 0) \\ = 1 & (h = 0) \end{matrix}$$

The function $C(h)$ is the autocorrelation of the sequence $\{y_r\}$, and the fact that $\{y_r\}$ has 'two-level' autocorrelation (0 and 1 in this case) is a third consequence of the randomness of $\{y_r\}$.

Regarding the sequence of 1s and -1s as a time-varying waveform $y(t)$, the autocorrelation function of $y(t)$ is defined as:

$$\phi(\tau) = \operatorname*{Lt}_{T\to\infty} \frac{1}{2T}\int_{-T}^{+T} y(t)y(t+\tau)dt.$$

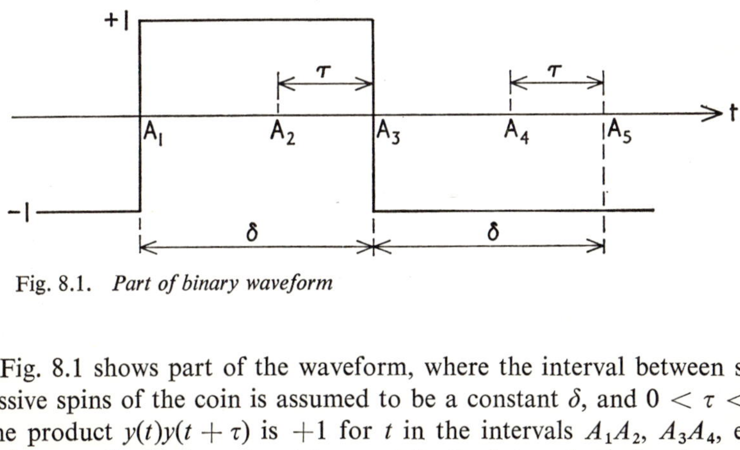

Fig. 8.1. *Part of binary waveform*

Fig. 8.1 shows part of the waveform, where the interval between successive spins of the coin is assumed to be a constant δ, and $0 < \tau < \delta$. The product $y(t)y(t + \tau)$ is $+1$ for t in the intervals A_1A_2, A_3A_4, etc., and is equally likely to be $+1$ as -1 in the intervals A_2A_3, A_4A_5, etc. Therefore the function $\phi(\tau)$, which is an average of the above product, equals $\dfrac{\delta - \tau}{\delta} = 1 - \dfrac{\tau}{\delta}$. By the same reasoning $\phi(\tau)$ comes to $1 + \dfrac{\tau}{\delta}$ for $-\delta < \tau < 0$. Combining these, therefore, $\phi(\tau) = 1 - \dfrac{|\tau|}{\delta}$ for $|\tau| < \delta$.

For $|\tau| > \delta$ there are no intervals of t over which the product is always $+1$. It is as likely to be $+1$ as -1 for all t, so $\phi(\tau) = 0$ for $|\tau| > \delta$. Since $\phi(\tau)$ is an average, its value does not depend on the position of the t-origin with respect to the waveform. Fig. 8.2 (*a*) shows the autocorrelation function of the coin-tossing waveform.

The reader may verify that the autocorrelation of a symmetrical square wave of period 2δ lying between the levels $+1$ and -1 is as shown in Fig. 8.2 (*b*).

The autocorrelation function of pure white noise $n(t)$ having an infinite spectrum is shown in Fig. 8.2 (*c*), where the height of the function at $\tau = 0$ is $\underset{T\to\infty}{\mathrm{Lt}} \frac{1}{2T}\int_{-T}^{+T}[n(t)]^2 dt$, namely the mean square value of the noise. The

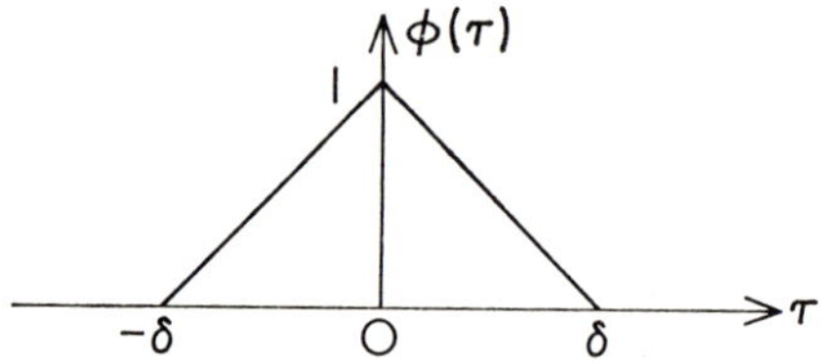

(*a*) *Random binary sequence*

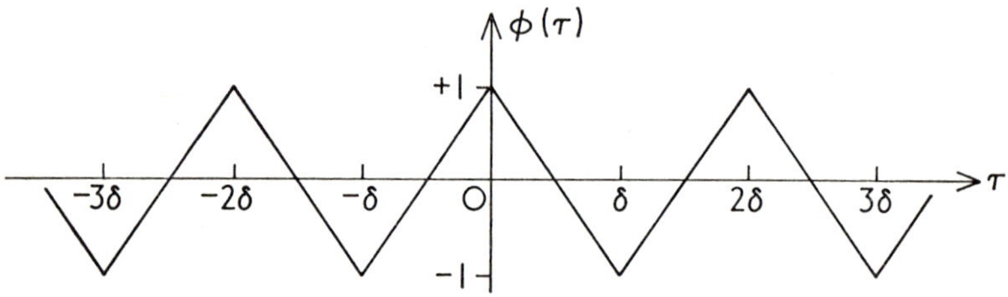

(*b*) *Symmetrical square wave*

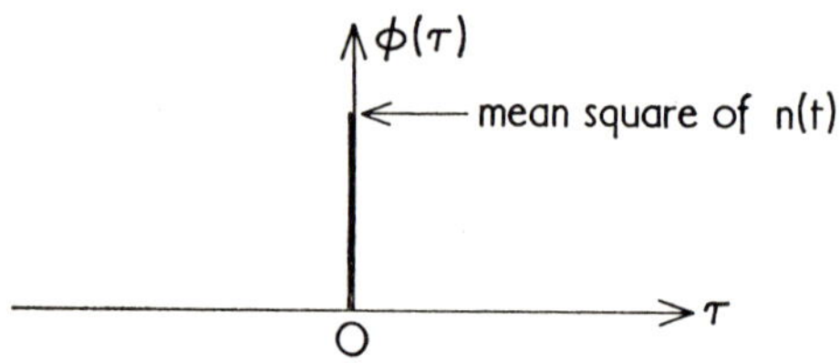

(*c*) *White noise*

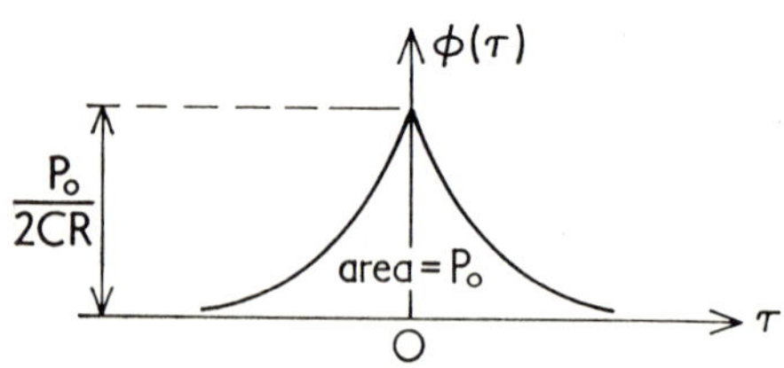

(*d*) *Filtered white noise*

Fig. 8.2. *Continuous autocorrelation functions*

autocorrelation of white noise is infinitely narrow in time τ since the instantaneous amplitude of the waveform can change suddenly at any time, so $n(t)$ and $n(t + \tau)$ are entirely uncorrelated for all τ other than zero. Although the coin-tossing waveform $y(t)$ also has an infinite spectrum it only changes suddenly at discrete instants of time, there being a finite range of values of τ over which $y(t)$ and $y(t + \tau)$ exhibit correlation. For this reason its autocorrelation function remains finite over a finite interval of τ.

The mean square value of a voltage or current waveform is also $\int_{-\infty}^{+\infty} P df$, where P is its power spectral density in watts/Hz. For white noise, since P is constant, either the mean square value is infinite or the power spectral density is zero. A more realistic source of 'white' noise is an ideal white-noise generator of finite power spectral density P_0 filtered by a simple CR network. The power spectral density of the real source is then $P_0/(1 + w^2C^2R^2)$. The Wiener–Khintchine formula relating the autocorrelation and power spectral density of a waveform therefore gives:

$$\begin{aligned}\phi(\tau) &= \frac{1}{2\pi}\int_{-\infty}^{+\infty} P(w)e^{jw\tau}dw \\ &= \frac{P_0}{2\pi}\int_{-\infty}^{+\infty} \frac{e^{jw\tau}}{1 + w^2C^2R^2}\, dw \\ &= \frac{P_0}{2CR}e^{-|\tau/CR|}.\end{aligned}$$

which is a pulse as shown in Fig. 8.2 (*d*) whose height (equal to the mean square value) increases and whose width decreases as CR is diminished. The area under this pulse is $\frac{P_0}{2CR}\int_{-\infty}^{+\infty} e^{-|\tau/CR|}d\tau = P_0$. Thus in the limit as $CR \to 0$ the autocorrelation of this 'white' noise becomes $P_0\delta(\tau)$, where P_0 is its power spectral density at low frequencies in watts/Hz and $\delta(\tau)$ is the Dirac delta function.

We have drawn a distinction between autocorrelation of a sequence $C(h)$ (discrete autocorrelation) and autocorrelation of a waveform $\phi(\tau)$ (continuous autocorrelation). For switching waveforms having square edges and fixed clock period, of which the coin-tossing and symmetrical square waves discussed above are examples, it is easily verified that the continuous autocorrelation function $\phi(\tau)$ is linear within any interval $h\delta < \tau < (h + 1)\delta$, where h is a whole number and δ is the clock period, and that $\phi(\tau)$ at the values $\tau = h\delta$ is the same as the discrete autocorrelation $C(h)$. Given $C(h)$, therefore, the form of $\phi(\tau)$ follows directly and consists of straight lines joining successive values of $C(h)$.

Randomness Properties of *m*-sequences

We now consider how far a binary *m*-sequence resembles the coin-tossing waveform. Consider an *m*-sequence of period $v = 2^n - 1$ generated by an *n*-stage maximal register.

(i) By writing successive *n*-bit patterns of the sequence, as shown below in a case with $n = 5$:

$$\begin{array}{cccccccccccc}
1 & 1 & 1 & 1 & 1 & & & & & & & \\
 & 1 & 1 & 1 & 1 & 0 & & & & & & \\
 & & 1 & 1 & 1 & 0 & 1 & & & & & \\
 & & & \cdot & & & & & & & & \\
 & & & & \cdot & & & & & & & \\
 & & & & & \cdot & & & & & & \\
 & & & & & & 0 & 0 & 0 & 1 & 1 & \\
 & & & & & & & 0 & 0 & 1 & 1 & 1 \\
 & & & & & & & & 0 & 1 & 1 & 1 & 1
\end{array}$$

and observing that all the non-zero *n*-bit numbers occur once in the case of *m*-sequences, we see that the number of 1s in a period is 2^{n-1} and the number of 0s is $2^{n-1} - 1$, i.e. the number of 1s exceeds the number of 0s by one in each period of $2^n - 1$ clock pulses. Thus the *m*-sequence differs slightly from the coin-tossing sequence in this respect, though the difference becomes less significant as *n* increases.

(ii) Defining a run of 1s as a sequence of consecutive 1s flanked at either end by a zero, and correspondingly for a run of 0s, since an *n*th-order *m*-sequence contains each *n*-bit non-zero pattern just once, each period contains: one run of *n* 1s, no run of $n - 1$ 1s, one run of $n - 2$ 1s, no run of *n* 0s, one run of $n - 1$ 0s, and one run of $n - 2$ 0s.

As regards runs of length *i* $(1 \leqslant i < n - 2)$ the configuration

$$\overbrace{\underbrace{011\ldots1}_{i}0xxx\ldots x}^{n} \quad \text{and} \quad \overbrace{\underbrace{100\ldots0}_{i}1xxx\ldots x}^{n},$$

each of which contains a run of length *i*, each occur 2^{n-i-2} times in a period since the $n - i - 2$ undefined digits can be completed in 2^{n-i-2} ways. This is therefore the number of runs (of *i*s or 0s) of length *i* in each period. Note that in completing the above configurations additional runs of length *i* may be formed, e.g. one way of completing $010xx$ is 01010, leading to a configuration containing two runs of length one. A little thought, however, shows that this does not invalidate the above enumeration.

Thus, except for lengths n and $n - 1$, the m-sequence is equivalent to the coin-tossing sequence as regards the number of runs of given length, there being twice as many runs of length one as of length two, twice as many of length two as of length three, etc.

(iii) In chapter six it was shown that if $\{y_r\}$ is an m-sequence of 1s and 0s, an integer l may be found for which $y_{r+l} = y_r \oplus y_{r+k}$ for all integral values of k other than 0, v, $2v$, etc., where v is the period of the sequence. If each 1 in the sequence is replaced by -1 and each 0 by $+1$ this relation becomes $y_{r+l} = y_r y_{r+k}$, where the terms are now either $+1$ or -1. For a periodic sequence of period v the discrete autocorrelation may be defined as $C(h) = \frac{1}{v}\sum_{r=1}^{v} y_r y_{r+h}$. In this case, therefore, for $h \not\equiv 0(\text{mod } v)$, $C(h) = \frac{1}{v}\sum_{r=1}^{v} y_{r+l} = -\frac{1}{v}$, since the number of -1s in one period of $\{y_r\}$ exceeds the number of $+1$s by one.

For $h \equiv 0(\text{mod } v)$, $C(h) = \frac{1}{v}\sum_{r=1}^{v}(1) = 1$. Thus the m-sequence has two-level autocorrelation, and resembles the coin-tossing sequence in this respect also, the resemblance increasing as v tends to infinity.

The autocorrelation function $\phi(\tau)$ of the corresponding time variation $y(t)$ is as before

$$\underset{T\to\infty}{\text{Lt}} \frac{1}{2T}\int_{-T}^{+T} y(t)y(t+\tau)dt$$

$$= \frac{1}{v\delta}\int_0^v y(t)y(t+\tau)dt,$$

since the sequence has a time period $v \times$ the clock period δ. From what has been said above about the relation between $C(h)$ and $\phi(\tau)$ for switching waveforms it follows that $\phi(\tau)$ has the form shown in Fig. 8.3.

Periodic sequences having two-level autocorrelation are called perfect sequences. If in addition such sequences possess property (i) above, namely the number of 1s and -1s (or 1s and 0s) in a period differs at

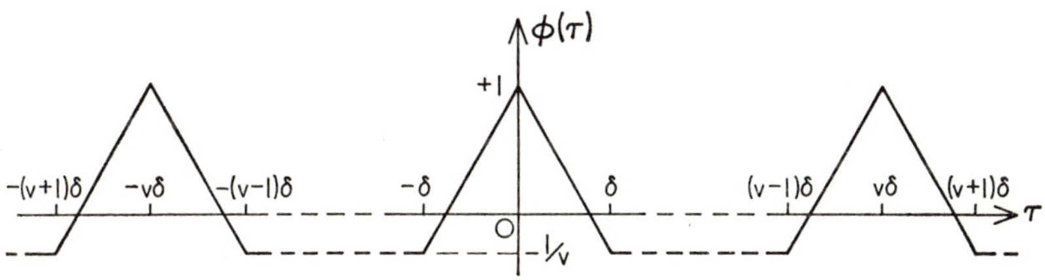

Fig. 8.3. *Autocorrelation function of an m-sequence*

most by one, they are called pseudo-random or pseudo-noise sequences. The binary *m*-sequence is therefore an example of a pseudo-random sequence. *M*-sequences are the only periodic sequences possessing in addition property (ii) above, namely the number of runs of length i in a period is twice the number of runs of length $i + 1$.

In the next chapter we examine other deterministic periodic sequences possessing one or more of the above three properties.

Cross-correlation

The cross-correlation between two waveforms $x(t)$ and $y(t)$ is defined as:

$$\phi_{xy}(\tau) = \underset{T\to\infty}{\text{Lt}} \frac{1}{2T}\int_{-T}^{+T} x(t)y(t+\tau)dt$$

(The alternative definition $\underset{T\to\infty}{\text{Lt}} \frac{1}{2T}\int_{0}^{T} x(t)y(t+\tau)dt$ is sometimes used.)

Autocorrelation is therefore a special case of cross-correlation, in which the two waveforms $x(t)$ and $y(t)$ are the same. Again, if $x(t)$ and $y(t)$ are waveforms corresponding to sequences $\{x_r\}$ and $\{y_r\}$ of 1s and -1s clocked by the same clock pulse train, we can define a discrete cross-correlation between the sequences as

$$C_{xy}(h) = \frac{1}{v}\sum_{r=1}^{v} x_r y_{r+h},$$

where v is the period of the combined sequence formed by the termwise product of the two sequences.

Two cases of special importance will be considered, the first in which $\{x_r\}$ and $\{y_r\}$ are binary sequences of relatively prime period v_1 and v_2, and the second in which $\{x_r\}$ and $\{y_r\}$ are binary *m*-sequences of equal period v.

1. Binary sequences of relatively prime period. In this case the period of the combined sequence is v_1v_2. Suppose $\{x_r\}$ and $\{y_r\}$ are the sequences $- - - + - + +$ repeats, and $- - +$ repeats, respectively (where $-$ and $+$ represent -1 and $+1$, respectively). For $h = 0$ we have

$\{x_r\} = - - - + - + + - - - + - + + - - - + - + +$, etc.
$\{y_r\} = - - + - - + - - + - - + - - + - - + - - +$, etc.
$\{x_r y_r\} = + + - - + + - + - + - - - - - + + + + - +$, repeats.

$C_{xy}(0)$ therefore equals $\frac{1}{21}(11 - 10) = \frac{1}{21}$, i.e.

$$\frac{\text{no. of agreements} - \text{no. of disagreements}}{\text{no. of agreements} + \text{no. of disagreements}}.$$

To obtain C_{xy} for values of h other than zero we only need to work it out for $h = 1$ and 2 since $\{y_{r+3}\} = \{y_r\}$, $\{y_{r+4}\} = \{y_{r+1}\}$, $\{y_{r+5}\} = \{y_{r+2}\}$, $\{y_{r+6}\} = \{y_r\}$, etc. It is however, not even necessary to do this, for since $\{x_r\}$ and $\{y_r\}$ take on all possible relative phase relations over one combined period of 21, shifting $\{y_r\}$ with respect to $\{x_r\}$ by some value h is equivalent to shifting both by some other value $h'(< 21)$. For example, for $h = 1$, $x_{r+7} = x_r$ and $y_{r+7} = y_{r+1}$ and so the correlation between $\{x_r\}$ and $\{y_{r+7}\}$ is the same as that between $\{x_{r+7}\}$ and $\{y_{r+7}\}$, i.e. between $\{x_r\}$ and $\{y_r\}$. For $h = 2$, $x_{r+14} = x_r$ and $y_{r+14} = y_{r+2}$, so the correlation between $\{x_r\}$ and $\{y_{r+2}\}$ is the same as that between $\{x_{r+14}\}$ and $\{y_{r+14}\}$, i.e. between $\{x_r\}$ and $\{y_r\}$. Consequently $C(1) = C(2) = C(0)$. In general, if $\{x_r\}$ and $\{y_r\}$ have relatively prime periods, the discrete cross-correlation between them is constant. Obviously the continuous cross-correlation between the corresponding waveforms $x(t)$ and $y(t)$ is likewise constant.

It is instructive to consider the above property in terms of probability. Sequences whose periods are relatively prime are 'statistically independent' since when juxtaposed they take on all possible relative phases. In this case, therefore, the joint probability $p(a, b)$ that x_r and y_r are a and b respectively is the product of the probability that $x_r = a$ and the probability that $y_r = b$. The cross-correlation between $\{x_r\}$ and $\{y_r\}$, which equals the probability that $x_r y_{r+k} = +1$ minus the probability that $x_r y_{r+k} = -1$, i.e. $[p(1, 1) + p(-1, -1)] - [p(1, -1) + p(-1, 1)]$, can therefore be calculated in terms of the probability of 1s and -1s in each sequence. In the above example

$$p_x(1) = \frac{3}{7},\; p_x(-1) = \frac{4}{7},\; p_y(1) = \frac{1}{3} \text{ and } p_y(-1) = \frac{2}{3},$$

so
$$C_{xy} = \left[\left(\frac{3}{7} \times \frac{1}{3}\right) + \left(\frac{4}{7} \times \frac{2}{3}\right)\right] - \left[\left(\frac{3}{7} \times \frac{2}{3}\right) + \left(\frac{4}{7} \times \frac{1}{3}\right)\right]$$

$$= \frac{1}{21}, \text{ as before.}$$

2. Binary *m*-sequences of equal period. If $\{x_r\}$ and $\{y_r\}$ are the sequences, the cross-correlation $\frac{1}{v} \sum_{r=1}^{v} x_r y_{r+h}$ takes on different values according to the value of h. No formula is known giving the number of different cross-correlations for any given pair of sequences, but it can be shown that this number cannot exceed the number of cosets associated with period v. For example with $v = 7$ the cosets are (7), (1, 2, 4) and (3, 6, 5), so the number cannot exceed 3. The case of $v = 15$ will be considered by way of

demonstration, the cosets being (15), (1, 2, 4, 8), (3, 6, 9, 12), (5, 10), and (7, 11, 13, 14) in this case.

There are two *m*-sequences of length 15. If one is $\{x_r\}$ the other is obtained by sampling $\{x_r\}$ every qth term, where q belongs to the last coset (i.e. q is prime to 15). Suppose the first term x_0 of $\{x_r\}$ is so chosen that $\{x_{2r}\}$ and $\{x_r\}$ are termwise identical, then the sequences $\{y_{2r}\}$ and $\{y_r\}$, where $y_r = x_{qr}$, will likewise be termwise identical. With this choice of starting-point the terms x_r, where r belongs to any one coset, are equal, and the terms y_r, where r belongs to any one coset, are equal.

If we write down all the terms in $vC_{xy}(h) = \sum_{r=0}^{v-1} x_r y_{r+h} = \sum_{r=0}^{v-1} x_r x_{q(r+h)}$, and group these as follows:

$$
\begin{aligned}
&(x_0 x_{qh}) + (x_1 x_{q(1+h)} + x_2 x_{q(2+h)} + x_4 x_{q(4+h)} + x_8 x_{q(8+h)}) + \\
&(x_3 x_{q(3+h)} + x_6 x_{q(6+h)} + x_9 x_{q(9+h)} + x_{12} x_{q(12+h)}) + \\
&(x_5 x_{q(5+h)} + x_{10} x_{q(10+h)}) + \\
&(x_7 x_{q(7+h)} + x_{11} x_{q(11+h)} + x_{13} x_{q(13+h)} + x_{14} x_{q(14+h)})
\end{aligned}
$$

we find, by virtue of the above equalities between the terms, that if h belongs to any one coset, for each value of h in the coset the corresponding brackets are equal. For example, with $q = 7$ and $h = 5$ the third bracket comes to $x_3 x_{56} x_6 x_{77} + x_9 x_{98} + x_{12} x_{119} = x_3(x_7 + x_1 + x_1 + x_7)$, and for $h = 10$ it comes to $x_3 x_{91} + x_6 x_{112} + x_9 x_{133} + x_{12} x_{154} = x_3(x_1 + x_7 + x_7 + x_1)$, which is the same.

In general $C_{xy}(h)$ depends, not on h, but on the coset to which h belongs, and so it can have no more values that there are cosets. It is noteworthy that this property holds for all sequences having multiplier cosets since it arises entirely from the term equalities dictated by the cosets.

Fourier Analysis of *m*-sequences

A knowledge of the frequency spectrum of sequences is necessary in order to determine the bandwidth needed for the undistorted transmission of a sequence-modulated carrier.

If $y(t)$ is a periodic function of period T it can be expressed as a Fourier series $\sum_{n=-\infty}^{+\infty} a_n e^{2\pi jnt/T}$, where a_n are the exponential Fourier coefficients given by:

$$a_n = \frac{1}{T}\int_0^T y(t) e^{-2\pi jnt/T}\, dt$$

In general a_n is a complex number whose modulus is one half the amplitude of the nth harmonic in the spectrum of $y(t)$. The DC level of $y(t)$ is a_0.

If $y(t)$ is a sequence of period $v\delta$ whose value is y_1 in the interval 0 to δ, y_2 in the interval δ to $2\delta, \ldots, y_v$ in the interval $(v - 1)\delta$ to $v\delta$, its spectrum consists of a DC component and components at frequencies $\frac{n}{v\delta}$ ($n = \pm 1, \pm 2, \pm 3, \ldots$). The exponential Fourier coefficients in this case are given by

$$a_n = \frac{1}{v\delta}\int_0^{v\delta} y(t)e^{-2\pi jnt/T}\, dt$$

$$= \frac{1}{v\delta}\sum_{r=1}^{v}\int_{(r-1)\delta}^{r\delta} y_r e^{-2\pi jnt/T}\, dt$$

$$= \frac{1}{v\delta}\sum_{r=1}^{v}\int_0^{\delta} y_r e^{-2\pi jn(t+(r-1)\delta)/v\delta}\, dt$$

$$= \frac{1}{v\delta}\sum_{r=1}^{v} y_r e^{-2\pi jn(r-1)/v}\int_0^{\delta} e^{-2\pi jnt/v\delta}\, dt$$

$$= \frac{1 - e^{-2\pi jn/v}}{2\pi jn}\sum_{r=1}^{v} y_r e^{-2\pi jn(r-1)/v}$$

$$= \frac{1}{v}e^{-\pi jn/v}\left[\frac{\sin\frac{n\pi}{v}}{\frac{n\pi}{v}}\right]\sum_{r=1}^{v} y_r e^{-2\pi jn(r-1)/v}$$

$$= \frac{1}{v}\left[\frac{\sin\frac{n\pi}{v}}{\frac{n\pi}{v}}\right] e^{\pi jn/v}\sum_{r=1}^{v} y_r e^{-2\pi jnr/v}$$

The conjugate of a_n is therefore given by

$$\bar{a}_n = \frac{1}{v}\left[\frac{\sin\frac{n\pi}{v}}{\frac{n\pi}{v}}\right] e^{-\pi jn/v}\sum_{s=1}^{v} y_s e^{2\pi jns/v}$$

Therefore

$$|a_n|^2 = a_n\bar{a}_n = \frac{1}{v^2}\left[\frac{\sin\frac{n\pi}{v}}{\frac{n\pi}{v}}\right]^2 \sum_{r=1}^{v} y_r e^{-2\pi jnr/v}\sum_{s=1}^{v} y_s e^{+2\pi jns/v}$$

After some manipulation, and remembering that $y_{r+v} = y_r$, this comes to $\frac{1}{v^2}[C(0) + (v - 1)C(1)]$ if $n = 0$, and

$$\frac{1}{v^2}\left[\frac{\sin\frac{n\pi}{v}}{\frac{n\pi}{v}}\right]^2 [C(0) - C(1)] \text{ if } v \dagger n,$$

in the case of a perfect sequence having two-level autocorrelation $C(0)$ and $C(1)$, where $C(h)$ is defined as before.

In particular, for a binary m-sequence of $+1$s and -1s for which $C(0) = 1$ and $C(1) = -\frac{1}{v}$ the above expressions come to

$$\frac{1}{v^2} \text{ and } \frac{v+1}{v^2}\left[\frac{\sin\frac{n\pi}{v}}{\frac{n\pi}{v}}\right]^2.$$

The first zero of

$$\left[\frac{\sin\frac{n\pi}{v}}{\frac{n\pi}{v}}\right]$$

occurs when $n = v$, i.e. at a frequency $\frac{n}{v\delta} = \frac{1}{\delta}$, so the effective bandwidth of a perfect sequence is determined only by the clock pulse interval, and not by the sequence at all.

The above results may also be obtained by using the Wiener–Khintchine formulae relating the power spectrum of a waveform to its autocorrelation function, $|a_n|^2$ being proportional to the power contained in the nth harmonic in the spectrum of $y(t)$—see refs. 6 and 7.

Waveforms Derivable from Maximal Shift Registers

So far we have considered the properties of binary waveforms, i.e. waveforms which assume one of two levels 0 and 1, or -1 and $+1$. We now examine the properties of waveforms formed by adding weighted versions of binary waveforms, since these are multi-level waveforms which under certain circumstances are noise-like. The subject is discussed by Davies.[8, 9]

Consider the arrangement of Fig. 8.4 in which $\{y_1\}, \{y_2\}, \ldots, \{y_N\}$ are statistically independent synchronously clocked random binary sequences, each of which assumes the value 0 or 1 with equal likelihood. Each sequence $\{y_r\}$ is weighted by a corresponding factor h_r and the output $\{y\}$ is the sum of the weighted sequences. Thus, between a pair of consecutive clock pulses $\{y\}$ can have any one of the values $0; h_1, \ldots, h_n; h_1 + h_2, \ldots, h_{N-1} + h_N; h_1 + h_2 + h_3, \ldots, h_{N-2} + h_{N-1} + h_N; \ldots; h_1 + h_2 + h_3 + \ldots + h_N$. Each of these has the same probability $1/2^N$ since it corresponds to one of the 2^N possible input configurations, each of which is equally likely.

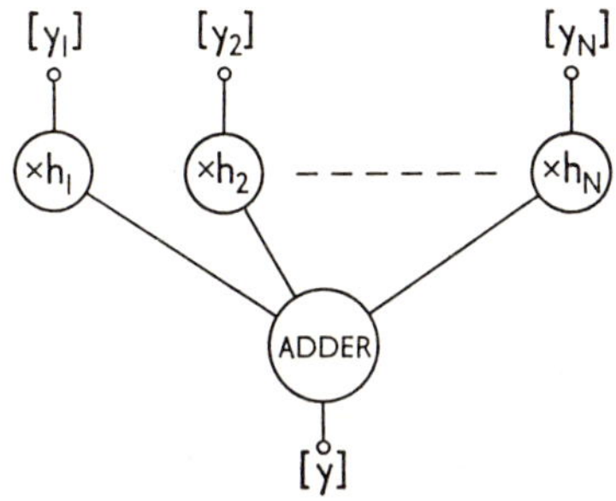

Fig. 8.4. *Sum of weighted binary sequences*

If a number K of these values are the same and equal to y, say, the probability that $\{y\}$ has the value y is $K/2^N$. This is exactly the coefficient of x^y in the polynomial $P(x) = \frac{1}{2^N}(1 + x^{h_1})(1 + x^{h_2}) \ldots (1 + x^{h_N}) = \frac{1}{2^N}[1 + (x^{h_1} + \ldots + x^{h_N}) + (x^{h_1+h_2} + \ldots + x^{h_{N-1}+h_N}) + (x^{h_1+h_2+h_3} + \ldots + x^{h_{N-2}+h_{N-1}+h_N}) + \ldots + (x^{h_1+h_2+\ldots+h_N})]$.

The polynomial $P(x)$ is therefore a convenient expression giving the probabilities of all possible values which $\{y\}$ can assume in terms of the coefficients of powers of x.

As examples consider the two cases $N = 6, h_1 = h_2 = h_3 = h_4 = h_5 = h_6 = 1$, and $N = 6, h_1 = h_2 = 1, h_3 = h_4 = 2$, and $h_5 = h_6 = 4$. $P(x)$ in the first case is $\frac{1}{2^6}(1 + x)^6 = \frac{1}{2^6}(1 + 6x + 15x^2 + 20x^3 + 15x^4 + 6x^5 + x^6)$, and so the probabilities that the output has the values 0, 1, 2, 3, 4, 5, or 6 are in the proportions 1 : 6 : 15 : 20 : 15 : 6 : 1. This weighting arrangement therefore yields an output whose amplitude has a binomial probability distribution. The polynomial in the second case is

$$\frac{1}{2^6}(1 + x)^2(1 + x^2)^2(1 + x^4)^2 = \frac{1}{2^6}(1 + 2x + 3x^2 + 4x^3 + 5x^4 + 6x^5 + 7x^6 + 8x^7 + 7x^8 + 6x^9 + 5x^{10} + 4x^{11} + 3x^{12} + 2x^{13} + x^{14}),$$

and so the output amplitude in this case has a triangular probability distribution.

The basis of the above derivation is that each of the 2^N possible configurations of input values is equally likely. If the inputs are the digits of an N-bit binary number which is clocked sequentially through all its 2^N values in some order, within each period of 2^N clock pulses each input configuration occurs just once, and so its 'probability' is $1/2^N$. Consequently the above derivation holds in this case too. However there is an important difference between the two cases. Whereas in the former case since the inputs are independent any transition in the value of $\{y\}$ can occur from one clock period to the next, in the latter case this is not so. For example, if the N-bit number is derived from the stages of a simple N-bit binary counter, it is impossible for it to change from, say, 111 . . . to 101 . . . in one clock period, or if it is derived from the stages of an N-stage full-length shift register, i.e. one which cycles through all the possible states including the all-zero state, it is impossible for the number to change from 00 . . . 0 to 11 . . . 1 in one clock period. Thus, although the amplitude probability distributions of the output in the two cases are the same, the output waveform in the latter case is far from noise-like. The closest approximation to a true noise generator would be if the inputs $\{y_r\}$ were separate m-sequences all of different length.

The spectral distribution of the output waveform is the sum of the spectral distributions of the inputs each weighted by its corresponding factor. When the inputs are different m-sequences the output spectrum may readily be deduced from the formula given in the previous section for the spectrum of an m-sequence. When, however, the inputs are derived from successive stages of an N-stage shift register clocked at frequency $1/\delta$ and producing a sequence of length v, the output spectrum is affected by the weighted summation process itself as follows:

Suppose the sequence produced at the output of the first stage, expressed as a Fourier series, is $\sum_{n=\infty}^{+\infty} a_n e^{jnw_0 t}$, where $w_0 = \dfrac{2\pi}{v\delta}$, the output is

$$\begin{aligned} &h_1 \sum_n a_n e^{jnw_0 t} + h_2 \sum_n a_n e^{jnw_0(t-\delta)} \\ &+ \ldots + h_N \sum_n a_n e^{jnw_0[t-(N-1)\delta]} \\ &= \sum_n a_n e^{jnw_0 t}(h_1 + h_2 e^{-jnw_0\delta} + \ldots + h_n e^{-(N-1)jnw_0\delta}). \end{aligned}$$

It may be observed, for example, that the component of frequency $\dfrac{nw_0}{2\pi}$ is missing if $(h_1 + h_2 u + h_3 u^2 + \ldots + h_N u^{N-1})$ is zero, where $u = e^{-jnw_0\delta}$. Thus with $h_1 = h_2 = 1$, $h_3 = h_4 = 2$, and $h_5 = h_6 = 4$ the polynomial

$1 + u + 2u^2 + 2u^3 + 4u^4 + 4u^5$ has $(1 + u)$ as a factor, and therefore one of its zeros occurs when $u = -1$, i.e. when $nw_0\delta = \pm\pi, \pm 3\pi, \pm 5\pi$, etc., or $n = \pm v/2, \pm 3v/2, \pm 5v/2$, etc. Had the weighting factors been the same but in the order $h_1 = h_6 = 1$, $h_2 = h_4 = 4$, and $h_3 = h_5 = 2$, say, the corresponding polynomial in u would not have contained the factor $(1 + u)$ and the spectral distribution of the output would therefore have been different.

Chapter nine

Other Forms of Binary Sequences

Sequences discussed up to now have mostly been of the *linear feedback shift register* form, the most important of these being the *m-sequence* by virtue of its almost ideal pseudo-random properties on the one hand, and its ease of generation on the other. In this chapter we consider two-level, or perfect, sequences in general and their properties, certain other sequences with practical applications, the full-length sequence and other non-linear feedback shift-register sequences.

Two-level Sequences

Let $\{y_r\}$ be a binary sequence of period v containing a $+$ 1s and b $-$ 1s. Therefore $a + b = v$. The discrete autocorrelation $C(h)$ is defined by $vC(h) = \sum_{r=1}^{v} y_r y_{r+h}$. For $h = 0$, $vC(0) = \sum_{r=1}^{v} y_r^2 = v$, and so $C(0) = 1$.

Now form the sum $\sum_{h=1}^{v} vC(h)$ in two ways. First, $\sum_{h=1}^{v} vC(h) = \sum_{h=1}^{v} \sum_{r=1}^{v} y_r y_{r+h}$ $= \sum_{r=1}^{v} \sum_{h=1}^{v} y_r y_{r+h} = \sum_{r=1}^{v} y_r \sum_{h=1}^{v} y_{r+h}$. Since $\{y_r\}$ is periodic, both these latter sums are equal to $a - b$, and so $\sum_{h=1}^{v} vC(h) = (a - b)^2$. Secondly,

$$\sum_{h=1}^{v} vC(h) = v[C(1) + \ldots + C(v-1)] + vC(v)$$
$$= v[C(1) + \ldots + C(v-1)] + v,$$

since $$C(v) = C(0).$$

If the sequence is two-level, $vC(h)$ for $h \neq 0$ is a constant w, say. Therefore $\sum_{h=1}^{v} vC(h) = w(v-1) + v$. Thus a necessary (though not sufficient) condition for $\{y_r\}$ to be two-level is that the equation $(a - b)^2 = (2a - v)^2 = w(v - 1) + v$ should be satisfied by integers, in which case $w = \dfrac{(2a - v)^2 - v}{v - 1}$. Note that the 1s and $-$1s may be interchanged without affecting the value of w.

The following special cases are of interest:

1. The difference between the number of 1s and -1s, $a \sim b$, is one. If the sequence is two-level $w = -1$ and the sequence is pseudo-random. *M*-sequences and two-level quadratic residue sequences (see below) are of this type, and it will be shown below that the period of such sequences must be odd and of the form $4n - 1$.
2. The number of 1s and -1s is the same. In this case the above equation cannot be satisfied in integers, and so no such sequence can ever be two-level.
3. The pulse waveform, in which a (or b) $= 1$. This is obviously always two-level, and the above equation shows that $w = v - 4$.
4. For a two-level sequence to have zero off-peak autocorrelation its period $(a + b)$ must be $(a \sim b)^2$.
5. For a two-level sequence to have $w = +1$, $(a \sim b)^2$ must be $2v - 1$.

Consider the evaluation of $\sum_{r=1}^{v} y_r y_{r+h}$ in more detail. Suppose as r goes through the values 1 to v, $y_r y_{r+h}$ is $(1 . 1)$ c times, $(-1 . -1)$ d times, $(1 . -1)$ e times, and $(-1 . 1)$ f times. By writing y_r and y_{r+h} as follows:

	← a →		← b →	
	← c →	← e →	← f →	← d →
$\{y_r\}$	1 1 . . . 1	1 1 . . . 1	-1 -1 . . . -1	-1 -1 . . . -1
$\{y_{r+h}\}$	1 1 . . . 1	-1 -1 . . . -1	1 1 . . . 1	-1 -1 . . . -1
	← v →			

we see that $c + e = a$ and $d + f = b$ from $\{y_r\}$ while $c + f = a$ and $d + e = b$ from $\{y_{r+h}\}$. Also $c + d + e + f = v$. Thus $f = e = a - c = b - d$, $v = c + d + 2e$, and $vC(h) = \sum_{r=1}^{v} y_r y_{r+h} = (c + d) - (e + f) - c + d - 2e = v - 4e$.

If the sequence is two-level $vC(h)$ is a constant w for $h \not\equiv 0$, and the parameters $c = a - \dfrac{v - w}{4}$, $d = b - \dfrac{v - w}{4}$ and $e = \dfrac{v - w}{4}$ are therefore independent of h.

We see, for $w = -1$, that for e to be an integer v must be of the form $4n - 1$. More generally, for any two-level binary sequence $v - w$ is a multiple of 4, and since $c + d = w + 2e = w + \dfrac{v - w}{2} = \dfrac{v + w}{2}$, $v + w$ must be even. This is another necessary condition for a sequence to be two-level. That this is further to the one derived above is seen by considering the case of $v = 9$ and $a = 6$. The first condition gives $w = 0$, but since $v + w$ is odd such a sequence cannot be two-level.

The terms c, d, and e for two-level sequences may be expressed in terms of v and a or b by using the formula for w derived above. Thus $c = \frac{a(a-1)}{v-1}$, $d = \frac{b(b-1)}{v-1}$, and $e = \frac{ab}{v-1}$.

Evidently if a sequence consisting of 1s and -1s is two-level so also is the sequence formed by replacing each 1 by the level p and each -1 by the level q. In this case $vC(0)$ becomes $ap^2 + bq^2$ and $vC(h)$ becomes $cp^2 + dq^2 + 2epq = \frac{(ap+bq)^2 - (ap^2+bq^2)}{v-1}$. In particular if $p = 0$ and $q = 1$, i.e. each period of the sequence consists of b 1s and a 0s, $vC(h) = b$, $(h \equiv 0)$ and $\frac{b(b-1)}{v-1}$, $(h \not\equiv 0)$. If, further, the sequence is pseudo-random, with the number of 1s exceeding the number of 0s by one, $b = \frac{v+1}{2}$, so $vC(h) = \frac{v+1}{2}(h \equiv 0)$ and $\frac{v+1}{4}(h \not\equiv 0)$.

Levels Giving Zero Off-peak Autocorrelation

Let each period of a pseudo-random sequence of period v consist of $\left(\frac{v+1}{2}\right)$ levels p and $\left(\frac{v-1}{2}\right)$ levels q. For the out-of-phase autocorrelation to be zero

$$\left[\left(\frac{v+1}{2}\right)p + \left(\frac{v-1}{2}\right)q\right]^2 - \left[\left(\frac{v+1}{2}\right)p^2 + \left(\frac{v-1}{2}\right)q^2\right] = 0$$

Let $\frac{q}{p} = \rho$ (for 'ratio'), then

$$[(v+1) + (v-1)\rho]^2 - 2(v+1) + (v-1)\rho^2 = 0.$$

Solving for ρ gives $\rho = [-(v+1) \pm 2\sqrt{(v+1)}]/(v-3)$.
With these values of ρ the in-phase autocorrelation becomes

$$\frac{1}{v}\left[\left(\frac{v+1}{2}\right)p^2 + \left(\frac{v-1}{2}\right)p^2\rho^2\right]$$

$$= \frac{(v+1)p^2}{v(v-3)^2}\left[v^2 - v + 2 \pm 2(v-1)\sqrt{(v+1)}\right].$$

For example, a two-level pseudo-random sequence of period 15, whose period consists of seven levels $+1$ (i.e. $q = 1$), and eight levels $-\frac{1}{2}$ (i.e. $p = -\frac{1}{2}$), has zero out-of-phase autocorrelation and in-phase autocorrelation $\frac{3}{5}$. This may be compared with a similar sequence whose period consists of seven 1s and eight -1s, which has out-of-phase autocorrelation $-\frac{1}{15}$ and unity in-phase autocorrelation.

Two-level sequences having zero out-of-phase autocorrelation are used in the evaluation of system parameters, and the biasing method described above is a way of producing them.

Number of Runs in a Period

The number of runs in a period of a sequence $\{y_r\}$ may be calculated as follows: Suppose the sequence consists of 1s and -1s. Beginning a period of $\{y_r\}$ between a 1 and a -1, the period contains an equal number of runs of 1s and -1s. Let this number be n. If below such a period of $\{y_r\}$ is written the corresponding period of $\{y_{r+1}\}$, termwise disagreement is seen to occur at the end of each run of 1s and each run of -1s in $\{y_r\}$. Thus $2n = 2e_{h=1}$. For a two-level sequence $e_{h=1} = e = \dfrac{ab}{v-1}$, and so $n = \dfrac{ab}{v-1}$. For a pseudo-random sequence this comes to $\dfrac{v+1}{4}$.

Relation Between Two-level Sequences, Hadamard Difference Sets, and Hadamard Designs

A difference set D, modulo an integer v, is a set of k integers $d_1, d_2, \ldots d_k$ having the property that all integers 1 to $v-1$ may be expressed as the difference between pairs of numbers in the set in exactly λ ways. Thus (0, 3, 4, 5) is a difference set (modulo 7) since

$$
\begin{aligned}
1 &\equiv 4-3 \equiv 5-4 \quad (\text{mod } 7)\\
2 &\equiv 5-3 \equiv 0-5 \quad \text{,,}\\
3 &\equiv 3-0 \equiv 0-4 \quad \text{,,}\\
4 &\equiv 0-3 \equiv 4-0 \quad \text{,,}\\
5 &\equiv 5-0 \equiv 3-5 \quad \text{,,}\\
6 &\equiv 4-5 \equiv 3-4 \quad \text{,,}
\end{aligned}
$$

In this case $\lambda = 2$.

Evidently a difference set is unaltered by adding a constant to each member of the set. Since the total number of differences associated with D is $\lambda(v-1)$ on the one hand, and the number of permutations of two numbers drawn from k numbers, namely $k(k-1)$, on the other, the parameters v, k, λ of D are related by the equation $\lambda(v-1) = k(k-1)$. The satisfaction of this equation in integers is a necessary, but not sufficient, condition for a difference set (modulo v) to exist.

If $\{y_r\}$ is a two-level sequence of period v consisting of a 1s and b -1s, the rs for which $y_r = 1$ form a difference set (modulo v) since, for all

$h \not\equiv 0 \pmod{v}$, there are exactly c values (as defined in the first section) of r for which y_r and y_{r+h} are both 1, implying that r and $r + h \pmod{v}$ are members of the set. The relation $c(v - 1) = a(a - 1)$ derived above shows that v, a, c correspond to the parameters v, k, λ of the set.

For the same reason the values of r for which $y_r = -1$ likewise form a difference set with parameters v, b, and d related by the equation $d(v - 1) = b(b - 1)$. Thus the existence of a difference set implies the existence of a two-level binary sequence, and vice versa. For example the m-sequence $-1 - 1 - 111 - 11$, repeats, corresponds to the difference set (0, 1, 2, 5) formed by the -1 terms, with parameters 7, 4, 2, and to the difference set (3, 4, 6) formed by the $+1$ terms, with parameters 7, 3, 1.

If each term of a difference set D of modulus v is multiplied by the same integer m another difference set having the same parameters is formed, providing m is prime to v. This follows since if $(d_i - d_j)$ runs through all the integers 1 to $v - 1$, as is the case, $md_i - md_j = m(d_i - d_j)$ likewise runs through the integers 1 to $v - 1$. If m is such that D and the new set are the same (except possibly for a constant difference between the corresponding terms) m is called a multiplier of D, and $1, m, m^2, \ldots \pmod{v}$ form a *multiplier set*. As might be expected, and as is easily shown (see, for example, Everett[10]) this group of numbers is also a multiplier set of the sequence formed from D.

The existence or otherwise of multipliers is an unresolved question. A theorem due to Hall[11] defines conditions to be satisfied for a multiplier to exist, but in fact no difference set has been found which does not have a multiplier, which in all known cases is a prime divisor of $k - \lambda$. For example, the m-sequences of period 15 correspond to a difference set having parameters $v = 15$, $k = 8$, and $\lambda = 4$. Thus $k - \lambda = 4$, and the prime divisor 2 is a multiplier of the sequence.

Regarding Hadamard, or balanced incomplete block, designs, the existence of a two-level sequence of period v also implies the existence of a *cyclic symmetrical* balanced incomplete block design of modulus v, and vice versa. The following is an example of a balanced incomplete block design:

0_1	0_2	0_3		
0_1	0_2		0_4	
0_1	0_2			0_5
0_1		0_3	0_4	
0_1		0_3		0_5
0_1			0_4	0_5
	0_2	0_3	0_4	
	0_2	0_3		0_5
	0_2		0_4	0_5
		0_3	0_4	0_5

0_1, 0_2, . . . 0_5 are objects of different type. The objects are disposed in the array so that

1. Each row (or 'set') contains the same number (3) of different objects.
2. Each column contains the same number (6) of objects.
3. Every pair of objects is to be found in one or other of the sets, and moreover occurs in the same number (3) of sets. Thus the pair $0_1 0_2$ occurs in sets 1, 2, and 3, $0_1 0_3$ in sets 1, 4, and 5, etc.

In general if v is the number of different types of object (i.e. the number of columns), s is the number of sets, k is the number of objects in each set, g is the number of sets in which an object of given type occurs, and λ is the number of sets in which a given pair of objects occurs, we have:

Number of objects in the design $= gv = ks$.

Number of object pairs in the design of which one member of the pair is a given object (e.g. 0_1) $= g(k - 1) = \lambda(v - 1)$.

A *symmetrical* design is one in which the number of sets s equals the number of object types v. It therefore comprises a square array, and the above two relations reduce to $\lambda(v - 1) = k(k - 1)$, corresponding exactly to the relation deduced for a difference set.

If a 1 is placed in each square of a symmetrical design containing an object and a 0 in each empty square, each row becomes a binary sequence of period v. If it is then possible to interchange the columns and to interchange the rows in such a way that the sequence in the $(i + 1)$th row is the sequence in the ith row shifted one place to the left, the design becomes a *cyclic* symmetrical balanced incomplete block design, and the sequence in any one row of this design is two-level. For example the following is a cyclic design with $v = 7$, $k = 4$, and $\lambda = 2$:

	Objects						
	1	1	1	0	0	1	0
	1	1	0	0	1	0	1
	1	0	0	1	0	1	1
Sets	0	0	1	0	1	1	1
	0	1	0	1	1	1	0
	1	0	1	1	1	0	0
	0	1	1	1	0	0	1

The process of forming the sum $\sum_{r=1}^{7} y_r y_{r+3}$, for instance, amounts to finding the sets which contain a 1 in their first and fourth cells (i.e. those sets which contain the object pair $0_1 0_4$). There are just two such sets (the third and the sixth), so this sum is 2. Since *any* given pair of 1s appears in just two sets, all the sums $\sum_{r=1}^{7} y_r y_{r+h}$ $(h \not\equiv 0)$ equal 2, and the sequence is therefore two-level, with

$$\begin{aligned} C(h) &= \tfrac{4}{7}(h \equiv 0) \\ &= \tfrac{2}{7}(h \not\equiv 0) \end{aligned}$$

The parameters v, k, λ of a cyclic design therefore correspond to the period of, the number of 1s in, and the value of $vC(h)$ $(h \not\equiv 0)$ for, the corresponding two-level sequence of 1s and 0s.

Known Forms of Two-level Sequences

The most important class of two-level sequence is the *m*-sequence by virtue of its ease of generation. However, many other forms of perfect sequence are known, some of which are also pseudo-random. The most numerous (in terms of period) are the quadratic residue sequences of prime period v, where v is of the form $4n - 1$. These are also pseudo-random.

The quadratic residues of a number v, called simply residues (mod v), are all numbers congruent (mod v) to squares. Thus (modulus 11), $1^2 \equiv 1$, $2^2 \equiv 4$, $3^2 \equiv 9$, $4^2 \equiv 5$, $5^2 \equiv 3$. Continuing, we find $6^2 \equiv 3$, $7^2 \equiv 5$, $8^2 \equiv 9$, $9^2 \equiv 4$, and $10^2 \equiv 1$, so the residues are 1, 3, 4, 5, and 9. The non-residues are 2, 6, 7, 8, and 10. The number of residues and non-residues (modulus an odd prime) is the same.

A quadratic residue sequence is formed by making $y_r = \pm 1$ when r is a residue and ∓ 1 when r is a non-residue. The first term y_0 may be either a 1 or a -1. Thus for $v = 11$ the following are all quadratic residue sequences:

$$
\begin{array}{ccccccccccc}
1 & 1 & -1 & 1 & 1 & 1 & -1 & -1 & -1 & 1 & -1 \\
1 & -1 & 1 & -1 & -1 & -1 & 1 & 1 & 1 & -1 & 1 \\
-1 & 1 & -1 & 1 & 1 & 1 & -1 & -1 & -1 & 1 & -1 \\
-1 & -1 & 1 & -1 & -1 & -1 & 1 & 1 & 1 & -1 & 1
\end{array}
$$

Since a residue congruent to n^2 is also congruent to $(v - n)^2$, reversing these sequences does not produce new sequences.

We now show that, for v equal to a prime of the form $4n - 1$, such sequences are pseudo-random.

Fermat's theorem $r^{v-1} \equiv 1 \pmod{v}$, where v is an odd prime, leads by factorization to $r^{(v-1)/2} \equiv \pm 1 \pmod{v}$, and r is a residue or non-residue according as the right-hand side is $+1$ or -1. Thus the product of two residues or the product of two non-residues is a residue, whereas the product of a residue and a non-residue is a non-residue. If the terms y_r and y_s of a quadratic residue sequence of prime period are both either 1 or -1 therefore, the term y_{rs} is 1, but if either y_r or y_s is -1, y_{rs} is -1. Generally therefore $y_r y_s = y_{rs}$ for such sequences.

The sum $\sum_{r=0}^{v-1} y_r y_{r+h} = y_0 y_h + y_0 y_{-h} + \sum_{\substack{r=1 \\ r \neq v-h}}^{v-1} y_r y_{r+h}$. Since v is prime, a number r' between 1 and $v - 1$ exists for all r such that $rr' \equiv 1 \pmod{v}$, so $y_{r+h} = y_{r+(rr')h} = y_{r(1+r'h)} = y_r y_{1+r'h}$.

Thus $y_r y_{r+h}$ becomes $y_r^2 y_{1+r'h} = y_{1+r'h}$, since $y_r^2 = 1$. It then follows that

$$
\sum_{\substack{r=1 \\ r \neq v-h}}^{v-1} y_{1+r'h} = \sum_{r'=2}^{v-1} y_{r'}
$$

$$
= \sum_{r'=1}^{v-1} y_{r'} - y_1 = -y_1,
$$

and so

$$
vC(h) = \sum_{r=0}^{v-1} y_r y_{r+h} = y_0 y_h (y_1 + y_{-1}) - y_1.
$$

This is independent of h for $y_1 = -y_{-1}$, which is always the case when v is of the form $4n - 1$.

Among quadratic residues (mod v) are *biquadratic* residues, *octic* residues, etc., i.e. numbers congruent to fourth powers, eighth powers, etc. For example $4^2 \equiv 2^4 \equiv 3 \pmod{13}$, and $10^2 \equiv 16^2 \equiv 2^8 \equiv 9$ (mod 13). It has been shown[12] that for defined forms of v biquadratic residue sequences, i.e. ones in which $y_r = 1$ when r is congruent to a fourth power and -1 otherwise, and octic residue sequences, i.e. ones in which $y_r = 1$ when r is congruent to an eighth power and -1 otherwise, are two-level (though not pseudo-random). Because of the restrictions on possible values of v, sequences of these forms are rare—only three periods less than 100 are possible, for example.

It can be shown[13] that if $\{x_r\}$ and $\{y_r\}$ are quadratic residue sequences of (prime) periods u and v respectively, with $v = u + 2$, the sequence $\{z_r\}$ defined as follows is pseudo-random:

$$z_r = +1 \text{ for } r \equiv 0 \pmod{v}$$
$$z_r = -1 \text{ for } r \equiv 0 \pmod{u} \text{ but } r \not\equiv 0 \pmod{v}$$
$$z_r = x_r y_r \text{ otherwise.}$$

Sequences of this form are known as twin-prime sequences. The first two occur for $u = 3$, $v = 5$, and $u = 5$, $v = 7$.

Another class of two-level sequences having prime period v of the for $4n^2 + 27$ has been discovered by Hall.[11] To form a sequence in this class select a primitive root μ of v which enables the congruence $t \equiv 1 \pmod{6}$ to be satisfied, where t satisfies $\mu^t \equiv 3 \pmod{v}$. For example with $v = 31$ (of the required form) the primitive root 3 of 31 leads to $t = 1$, which satisfies $t \equiv 1 \pmod{6}$. The Hall sequence is then defined as follows:

$$y_r = 1 \text{ for } r = \mu^s \pmod{v}, \text{ where } s = 0, 1 \text{ or } 3$$
$$y_r = 0 \text{ otherwise}$$

Finally, it has been shown[14, 15] that if the non-zero terms of a p-nary m-sequence of period $p^n - 1$ are replaced by 0 and the zero terms by 1, the resulting binary sequence has period $\frac{p^n - 1}{p - 1}$ and is two-level. For example the third-order ternary m-sequence (of period $3^3 - 1 = 26$) 121120111002 02122102220010, repeats, leads to the two-level binary sequence of period thirteen 0000010001101, repeats.

A Distance Property of Pseudo-random Sequences[16]

The Hamming Distance between two words each consisting of v binary digits is the number of digits of one of them which has to be changed to make it the same as the other. If $\{y_r\}$ is a sequence of period v consisting of a + 1s and b − 1s and we define a dictionary consisting of the v words $(y_1 y_2 y_3 \ldots y_v)$, $(y_2 y_3 \ldots y_v y_1)$, . . ., $(y_v y_1 y_2 \ldots y_{v-1})$, the least distance between a pair of these words is maximized by making $\{y_r\}$ pseudo-random.

Since the words are cyclic shifts of one another the distance D_h between a pair of words is the number of −1s in the sum $vC(h) = \sum_{r=1}^{v} y_r y_{r+h}$. The number of $+1$s in the sum is therefore $v - D_h$, and since $vC(h) = (v - D_h) - D_h, D_h = \frac{1}{2}v[1 - C(h)]$.

If D is the least value of D_h,

$$2(v - 1)D \leqslant 2 \sum_{h=1}^{v-1} D_h \sum_{h=1}^{v-1} v[1 - C(h)].$$

We have shown above that $\sum_{h=1}^{v} vC(h) = (a - b)^2$, so $2(v - 1)D \leqslant v^2 - (a - b)^2$. For v odd, $(a - b)^2 \geqslant 1$, and so $2(v - 1)D \leqslant v^2 - 1$, i.e. $D \leqslant \frac{1}{2}(v + 1)$. But $\frac{1}{2}(v + 1)$ is the distance between any pair of words if $\{y_r\}$ is pseudo-random, since in this case $vC(h) = -1$ $(h \neq 0)$ in the expression for D_h.

Thus the pseudo-random sequence forms an appropriate basis for constructing a dictionary for use in a digital communication system since the words can tolerate, on average, the greatest number of digit errors before confusion arises between them. In fact it can be shown that the probability of incorrectly identifying a word from a dictionary consisting of a cyclically permuted sequence is least when the sequence is pseudo-random.

Barker Sequences

An important class of non-periodic sequence has been described by Barker.[17] A Barker sequence may be defined as a sequence $y_1y_2 \ldots y_v$ of length (not period) v consisting of $+1$s and -1s, flanked at either end either by zero or by completely uncorrelated sequences. The terms of the Barker sequence satisfy the condition $|C_b(h)| \leqslant 1$ $(h \neq 0)$, where $C_b(h) = \sum_{r=1}^{v-h} y_r y_{r+h}$. The former situation arises when the dynamics of a system are evaluated by applying a finite sequence to its input and examining its output (see chapter ten). The latter situation arises in digital communication systems where an information-carrying sequence is punctuated at regular intervals by a Barker sequence for purposes of synchronization.[18] In either case the requirement is that the contribution $C_b(h)$ of the Barker sequence to the total autocorrelation of the whole sequence (of which the Barker sequence forms a part) should be small. The condition above, in fact, ensures that the magnitude of this contribution is equal to or less than unity.

Barker sequences of lengths only up to 13 are known and so, despite their attractive properties, their application is limited. The table below gives values of $C_b(h)$ in the case of the Barker sequence . . . $+++++--++-+-+$ . . . of length 13:

h	0	1	2	3	4	5	6	7	8	9	10	11	12
$C_b(h)$	13	0	1	0	1	0	1	0	1	0	1	0	1

Evidently, since $y_r y_{r+h}$ is ± 1, $C_b(h)$ must be 0 or ± 1 according as $v - h$ is even or odd. The useful sequences are those for which $C_b(h)$, when not zero, is always $+1$ or always -1. It has been noted by Douce[30] that the periodic sequence formed by repeating a Barker sequence is always two-level.

Inverse Repeat Sequences

One of the main applications of binary sequences is in the evaluation of system dynamics. The usual procedure is to use the sequence as an input signal and to derive the impulse response of the system by cross-correlating this signal with the system output signal. For linear systems free of drift a binary sequence with two-level autocorrelation having an appropriate period relative to the settling time of the system is suitable. However, for the analysis of non-linear systems the higher order autocorrelations of the input signal have to be known, i.e. the values of the sums

$$C(h_1, h_2, \ldots, h_g) = \frac{1}{v} \sum_{r=0}^{v-1} y_r y_{r+h_1} y_{r+h_2} \cdots y_{r+h_g}$$

where v is the sequence period and $g + 1$ is the order of the autocorrelation. In certain respects the most desirable sequence for this purpose is one whose out-of-phase even-order autocorrelations are constant, i.e. independent of the phase shifts $h_1, h_2, \ldots, h_g$, and whose out-of-phase odd-order autocorrelations are zero. A fairly good approximation to such a sequence is the inverse repeat sequence, i.e. a sequence of $+1$s and -1s formed by inverting every other term of an m-sequence. An example of an inverse repeat sequence derived from the m-sequence $- - - + + - +$, repeats, is $|- + - - + + +|+ - + + - - -|$, repeats.

Since the period of an m-sequence is odd, the period of the inverse repeat sequence derived from it is twice the period of the m-sequence and the second half of the inverse repeat sequence cycle is the complement (term by term) of the first half. Such a sequence is therefore anti-symmetric, and by virtue of the property $y_{r+v/2} = -y_r$ the odd-order autocorrelations all vanish.

Simpson[19] has studied the statistics of inverse repeat sequences and has shown that the second-order autocorrelations of an inverse repeat sequence and of the m-sequence from which it is derived are equal or opposite according as h is even or odd. For most values of (h_1, h_2, h_3) the fourth-order autocorrelation of an inverse repeat sequence is either $\frac{2}{v}$ or $-\frac{2}{v}$, and for the remaining values it is ± 1. These irregularities, in the form of a ripple and occasional 'spikes', detract somewhat from the precision of the estimated system impulse response.*

Brown[20] has shown that antisymmetric sequences, defined by $y_{r+v/2} = -y_r$, used as test sequences provide a means of eliminating the effect of system drift on the estimate of the system's impulse response. The subject of drift, which has received a great deal of attention in the above context, will be considered in more detail in the next chapter. The same author[21] has recently defined a new class of binary sequence, in some ways related to antisymmetric sequences, which he calls 'maximally orthogonal'. Sequences in this class have the following autocorrelation:

$$
\begin{aligned}
C(h) = \frac{1}{v}\sum_{r=1}^{v} y_r y_{r+h} &= 1(h \equiv 0) \\
&= \frac{4-v}{v}\left(h \equiv \frac{v}{2}\right) \\
&= 0, \text{ otherwise.}
\end{aligned}
$$

Apart from their zero off-peak autocorrelation, such sequences have the advantage that the cross-correlation between sequences whose periods are harmonically related (i.e. $v_2 = 2^k v_1$) is small. Goodwin[22] describes an application of maximally orthogonal sequences to system identification.

Periodic Sequence Generation Using Shift Registers

Any prescribed sequence of period v can be produced directly by a shift register with appropriate feedback logic. Since 2^n is the greatest period derivable from an n-stage register, a necessary condition for such a register to produce the sequence is that $2^n \geqslant v$. Since a given register state always gives rise to the same feedback digit, however, this is not a sufficient condition. It is also necessary for n to be one more than the length of the

* Strangely enough, the use of a Hadamard design for the construction of sequences having constant out-of-phase autocorrelations of higher order than two does not seem to have been suggested. It is perfectly feasible to add to the requirements of a block design quoted above the additional constraints:

4. Every combination of three objects is to be found in one or other of the sets, and moreover occurs in the same number, μ, of sets.
5. Every combination of four objects is to be found in one or other of the sets, and moreover occurs in the same number, ν, of sets, etc.

It readily follows that $\lambda(k-2) = \mu(v-2)$, $\mu(k-3) = \nu(v-3)$, etc. Thus for a symmetrical design with $v = 4$, $k = 3$, $\lambda = 2$, the value $\mu = 1$ satisfies the first of these equations, whereas no integral value of μ exists for $v = 7, k = 4$, and $\lambda = 2$.

longest repeated pattern of digits in one period, the longest repeated pattern being of length k if in a period one finds either $x_1 \boxed{x_1x_2 \ldots x_k} x_k$ *and* $\bar{x}_1 \boxed{x_1x_2 \ldots x_k} \bar{x}_k$, or $x_1 \boxed{x_1x_2 \ldots x_k} \bar{x}_k$ *and* $\bar{x}_1 \boxed{x_1x_2 \ldots x_k} x_k$. Thus the sequence 1110010, repeats, can be generated by a three-stage register since $2^3 > 7$ and the pattern $\boxed{11}$ appears twice, once as a $0 \boxed{11} 1$ and once as $1 \boxed{11} 0$. On the other hand the quadratic residue sequence 11011100010, repeats, though only of period 11 requires a five-stage register (with non-linear feedback), since the longest repeated pattern is $\boxed{1011}$, appearing once as $1 \boxed{1011} 1$ and once as $0 \boxed{1011} 0$.

Shift-register generation of prescribed sequences is probably the most straightforward method. Having determined the number of stages needed it is only necessary to write down a truth table giving the feedback digit as a function of the successive register states, and to evolve therefrom the, feedback logic satisfying the table. For example, the truth table for the above quadratic residue sequence is:

y_{r-5}	y_{r-4}	y_{r-3}	y_{r-2}	y_{r-1}	y_r
1	1	0	1	1	1
1	0	1	1	1	0
0	1	1	1	0	0
1	1	1	0	0	0
1	1	0	0	0	1
1	0	0	0	1	0
0	0	0	1	0	1
0	0	1	0	1	1
0	1	0	1	1	0
1	0	1	1	0	1
0	1	1	0	1	1

which yields the feedback logic $y_r = \bar{y}_{r-4}\bar{y}_{r-1} + \bar{y}_{r-5}\bar{y}_{r-2} + y_{r-5}y_{r-4}\bar{y}_{r-3}$. Note that the truth table is not complete. A five-stage register has 32 states, and this logic gives the values of y_r for all the 32 states. The corresponding register therefore produces a number of sequences, though it happens that the above logic is such that the required sequence is always reached irrespective of the initial state of the register. Using the Karnaugh Map technique and exploiting the 21 'don't care' terms many different feedback logics could be evolved to give the same sequence, and associated with each logic would be a different state transition diagram.

A prescribed sequence of period v, however, may always be generated indirectly from an n-stage register, where n is the least integer making $2^n \geqslant v$, by arranging that the register produces a sequence (not the required one) of length v using the method described in chapter six, and including pick-off logic which converts the sequence actually generated into the required sequence. The pick-off logic will in general derive its inputs from the outputs of the n stage of the register, as shown in Fig. 9.1. The design of the pick-off, or sequence conversion, logic is accomplished using the usual truth-table procedure. Using this method the quadratic residue sequence above can be derived from a four-stage rather than a five-stage register.

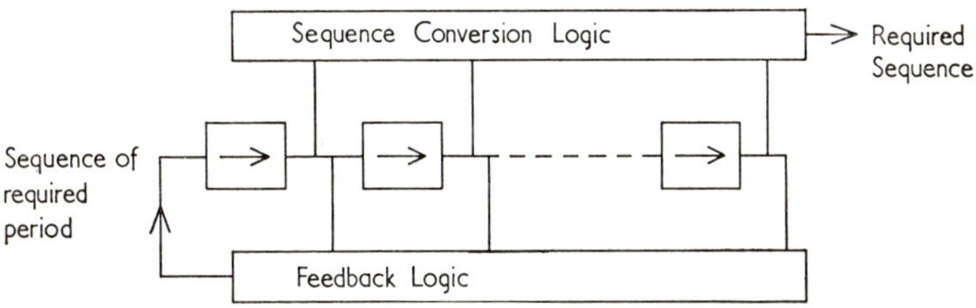

Fig. 9.1. *Generation of prescribed sequences using pick-off logic*

It is worth mentioning that special purpose logical systems for producing specific sequences have been described which are not of the feedback shift register form. For example, Darnell[23] describes a 'counter-de-code' system producing a quadratic residue sequence of period 19. The principle is claimed to be more economical in logic than the shift-register method for long sequences.

The Full-length Sequence

Although it has been suggested, perhaps fancifully, that the full-length binary sequence was known to the Indians many centuries ago on the grounds that the accents in a Hindu sentence of ancient origin represent the 1s in a full-length sequence of length 32, and the intervening syllables represent 0s, the first paper, to the author's knowledge, to appear on the now enormous subject of sequences was one by Martin[24] in 1934. This paper describes a systematic procedure for constructing two full-length binary sequences of order n, i.e. sequences of length 2^n containing in one period all the n-bit numbers 0 to $2^n - 1$. It may be said, therefore, that Martin proved the existence of full-length sequences of any order.

The principle of Martin's method is to start with either $(n - 1)$ 0s and use a 'preference rule' (1, 0), or with $(n - 1)$ 1s and use a preference rule (0, 1). Thus in the former case a 1 is added to the sequence as long as the latest n-bit pattern so formed has not previously occurred. If it has previously occurred a 0 is added instead. The procedure is continued until it is not possible to add either a 1 or a 0 without repeating a pattern. Omitting the first (or the last) $n - 1$ digits, the sequence which remains comprises one period of a full-length sequence of order n. A second such sequence results from the latter procedure. The following are the two sequences of order 3 formed by the above method: 0011101000, 1100010111

Considering the first procedure, the proof that all n-bit patterns occur in the sequence is as follows:

Except for the initial pattern any $(n - 1)$-bit pattern can only appear twice, once preceded by a 0 and once by a 1. On the first occasion it is followed by a 1 and on the second by a 0, and in each case a new n-bit pattern is formed. The only $(n - 1)$-bit pattern which can appear more than twice is the initial one since on its first appearance it has no preceding digit. The procedure of adding digits therefore stops on the third appearance of this pattern.

Except when a 0 has to be added after a pattern of $n - 1$ 0s the preference rule (1, 0) ensures that each n-bit pattern formed is non-zero. To ensure that all the non-zero n-bit patterns are formed, therefore, it is only necessary to make the procedure stop when the pattern of $n - 1$ 0s appears for the third time. This is accomplished by starting with a pattern of $n - 1$ 0s. The sequence then has the form:

$$\overbrace{0\,0 - - - 0}^{\text{1st appearance}}\,1\,X\,X - - - - - X\,X\,\overbrace{0\,\underbrace{0 - - - 0}_{\text{3rd appearance}}}^{\text{2nd appearance}}\!\!\!\!\underbrace{\,0}_{}\text{, stops}$$

We see that the all-zero n-bit pattern is included at the end of the sequence, and all the non-zero patterns occur between the first and last digits.

A modification of the above procedure, which enables all the full-length sequences of order n to be formed, is to start with any legitimate pattern of any length and proceed as above according to one or other of the preference rules. At some stage the left-hand $(n - 1)$-bit pattern will reappear for the third time at the right-hand end of the sequence, thereby preventing continuation. When this happens the first (i.e. LH) digit of the sequence is deleted, thereby allowing a digit to be added to the right-hand end. This process is continued until (in the case of the (1, 0) preference rule) the pattern of n 0s is reached, by which time all the non-zero n-bit patterns have been formed (for the reason given above, namely that any sequence formed using the (1, 0) rule which begins with $n - 1$ zeros and ends with *n* zeros must contain all the non-zero patterns). The following is an example for $n = 4$ using the initial pattern 011001011. The sequence forms as follows:

∅ 1̸ 1̸ ∅ ∅ 1̸ ∅ 1̸ 1̸ 1̸ 1̸ ∅ 1̸ 0 0 0   1 1 0 0 1 0 1 1 1 1 0 1 0 0 0 0

(first nine digits: initial pattern; last sixteen digits: full-length sequence)

As to the number of full-length sequences of order *n*, Posthumus, by what can only be described as a prodigious feat of diligence, actually constructed all the full-length sequences of order 1, 2, 3, 4, and 5 and found one of order one, one of order two, two of order three, sixteen of order four, and 2048 of order five (each of length 32)! He therefore conjectured that the number of order *n* was $2^{2^{n-1}-n}$.

De Brujn[25] proved the conjecture correct by showing that the ratio of the number of 'complete visits' in the complete state diagrams (see chapter two) of order $n + 1$ and *n* is $2^{2^{n-1}-1}$, a complete visit being a round trip from one of the states which passes through every other state just once, using roads between the states in the right direction. Each complete visit clearly corresponds to a different full-length sequence.

A complete visit must not be confused with a complete walk, which is a round trip from one of the states in which every road is used just once in the prescribed direction. The relation between the number of complete walks and complete visits is obtained by a process of 'doubling'. Consider the complete state diagram for a two-stage shift register shown in Fig. 9.2 (*a*).

This has four states and corresponds to the one full-length sequence 0011, repeats. Each of the eight roads may be defined by three digits, the last two being the digits defining the state which the road leaves and the first being the feedback digit giving rise to the state the road ends at.

The diagram is 'doubled' by regarding each road as a state. Two such states A and B are then joined by a road leading from A to B if the road corresponding to A in the original diagram leads directly through a single state into the road corresponding to B. In this way the complete state diagram for a three-stage shift register, as shown in Fig. 9.2 (*b*), is formed. It is then obvious that the number of complete visits (i.e. full-length sequences) for the three-stage diagram is the same as the number of complete walks round the two-stage diagram.

De Brujn then considers a net with *m* junctions and 2*m* one-way roads and its doubled version consisting of 2*m* junctions and 4*m* one-way roads, where each junction is the start of two roads and the finish of two roads. Figs. 9.2 (*a*) and (*b*) therefore represent the case with $m = 4$. By an argument of great beauty and delicacy he shows that the number of

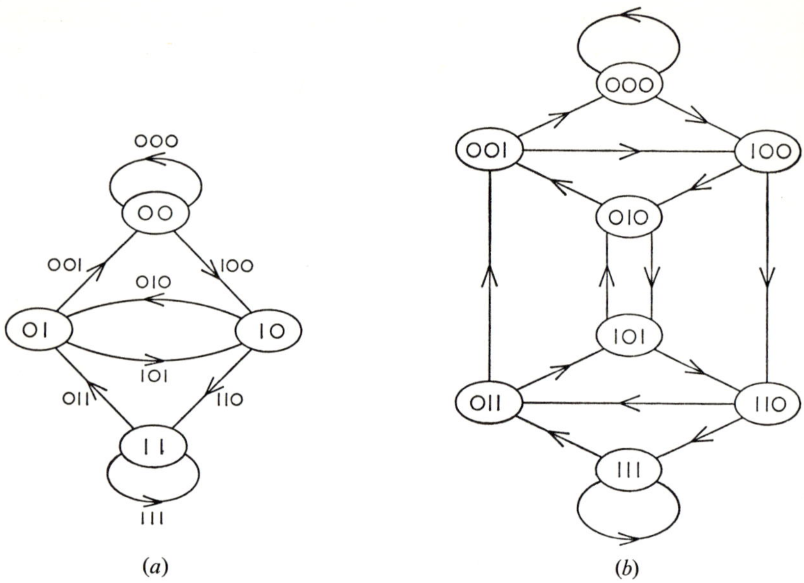

Fig. 9.2. *Complete state diagrams of order two and three*

complete walks round the doubled net is 2^{m-1} times the number of complete walks round the original net, from which the result follows at once.

It is interesting to note that a similar analysis is used in evaluating the number of Gray codes of given order, since a Gray sequence can likewise be represented as a net of junctions (states) and roads.

While the actual construction of all the full-length sequences of any order is implicit in de Brujn's proof, it is not at all simple to implement it, and Golomb[26] enunciates two theorems, the Key Word Theorem and the Key Sequence Theorem, which provide a direct method for constructing full-length sequences. The first of these theorems, which is an extension of Martin's preference rule method, produces 2^n sequences of order n which are not necessarily all distinct.

The second theorem enables full-length sequences of order n to be derived from a given full-length sequence of order $n - 1$ by the application of the preference rule $(P^+, \bar{P}^+)$, starting with an n-digit word whose first $n - 1$ digits are the initial digits of the sequence of order $n - 1$. In this case P^+ is not fixed, as in Martin's method, but is a function of the $n - 1$ digits preceding the digit being formed, i.e. when forming digit x_n, P^+ is a function of $x_1, x_2, \ldots, x_{n-1}$. The value of P^+ is determined by the feedback formula $x_1 \oplus F(x_2, \ldots, x_{n-1})$ defining the full-length sequence of order $n - 1$ and the corresponding preference rule $(P, \bar{P})$ used for forming this sequence. In tabulating P^+ for all possible combinations of $n - 1$ digits, therefore, it is first necessary to tabulate, from the given sequence of order $n - 1$, P for all combinations of $n - 2$ digits, and F for all combinations of $n - 1$ digits. $P^+(x_1, \ldots, x_{n-1})$ is then given by

$$0 = [P^+ \oplus 1 \oplus x_1 \oplus F(x_2, \ldots, x_{n-1})][P(x_2, \ldots, x_{n-1}) \oplus x_1 \oplus F(x_2, \ldots, x_{n-1})]$$

This formula leaves P^+ arbitrary for those combinations of $x_1, \ldots, x_{n-1}$ which make the right-hand bracket zero, and a different full-length sequence results for each possible P^+. Golomb shows that, given all full-length sequences of order $n - 1$ and all possible P^+, all the full-length sequences of order n are thereby formed, and the enumeration of these therefore follows.

It was shown above that a full-length sequence cannot be two-level. It may be noted, however, that since all n-bit patterns occur once in one period of an nth-order full-length sequence, and since in this set of patterns the first and the ith component agree and disagree the same number of times, for a full-length sequence of order n consisting of 1s and -1s,

$$\sum_{r=1}^{2^n} y_r y_{r+h} = 0 \text{ for } h = \pm 1, \pm 2, \ldots, \pm(n - 1).$$

Non-linear Feedback Shift Register Sequences

So far, apart from some discussion of word detection and certain special purpose sequences, the whole of this book has been concerned with linear sequences, i.e. sequences generated by shift registers with linear (modulo-2) feedback. This is because the theory of such sequences is straightforward and complete, and because many applications use m-sequences. If a linear sequence is described in terms of a polynomial there is no doubt about the primality or otherwise of the polynomial. If the polynomial is not prime its factorization is unique, and the associated cycle structure is very simply related to the factor structure. None of this is true in the case of non-linear sequences. Despite some recent attempts[27, 28, 29] to systematize their properties, non-linear sequences and sequence generators are still largely a mystery, and only the salient facts will be given here, most of which are fully explained by Golomb.[26]

In chapter two it was shown that for a feedback shift register to generate separate cycles the equation $y_r = F(y_{r-1}, \ldots, y_{r-n})$ for the feedback digit has to be of the form $y_r = y_{r-n} \oplus F_1(y_{r-1}, \ldots, y_{r-n+1})$. If y_r is described by a truth table of the following form, therefore:

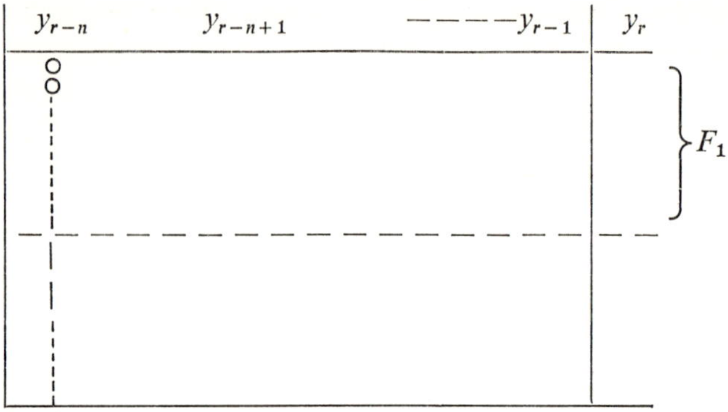

the top half of the y_r column must be the complement of the bottom half. In this case the top half of the truth table, excluding the y_{r-n} column represents the truth table for F_1.

For a cycling register the top half of the y_r column consists of zeros and so F_1 is zero, and the bottom half consists of ones. It was shown in chapter three that the number of cycles $Z(n)$ in this case is even (except for $n = 2$). Golomb shows that, for any shift register producing separate cycles, the effect on the number of cycles of altering one entry in the top half of the y_r column and the corresponding entry in the bottom half is to either increase or decrease the number of cycles by one. Beginning with the truth table for a cycling register, therefore, the number of cycles produced by a shift register generating separate cycles is even or odd according as the number of 1s in the top half of the y_r column (i.e. in F_1) is even or odd.

If the F_1 truth table is written as:

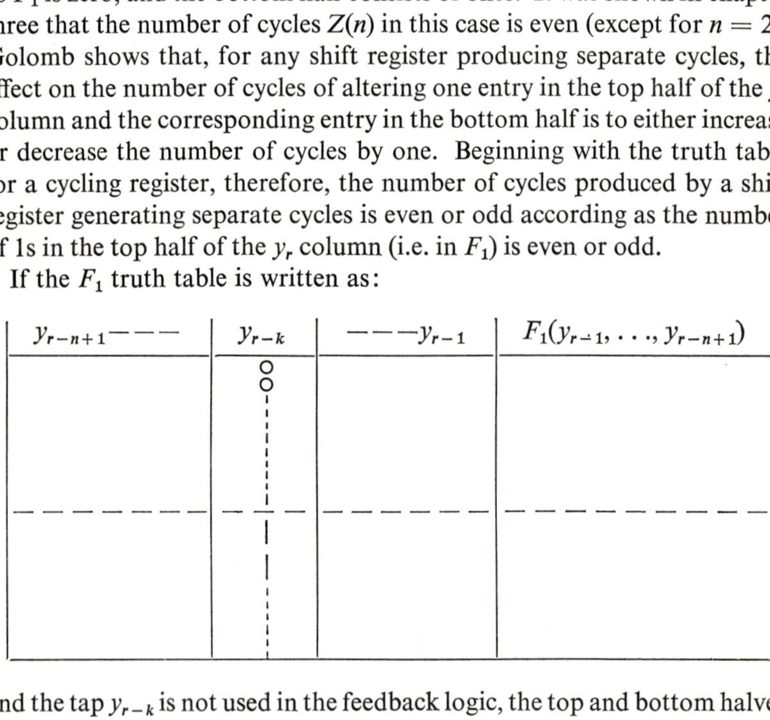

and the tap y_{r-k} is not used in the feedback logic, the top and bottom halves of the F_1 column are the same, and so the number of ones in the F_1 column is even. Unless y_{r-n} and all the other $n - 1$ taps are used in the feedback logic, therefore, the number of cycles generated is even. It follows that to generate a full-length sequence of length 2^n (i.e. a single cycle) from an n-stage register, all the n taps must be used.

In the case of linear feedback, since each column in the F_1 truth table contains an equal number of 1s and 0s and the modulo-two sum of any number of these columns produces a column with an equal number of 1s and 0s in it, the F_1 column necessarily contains the same number of 1s and 0s. This number is therefore even, and so the number of cycles produced by a linear feedback shift register is always even.

An F_1 truth table whose output column contains the same number of 1s and 0s is said to be 'balanced'. A shift register generating separate cycles whose F_1 truth table is balanced therefore generates an even number of cycles. A *non-linear* register of this kind must involve at least four taps, namely y_{r-n} and *three* (or more) from the stage outputs $y_{r-1}, \ldots, y_{r-n+1}$. Two would not be sufficient since any non-linear combination of two variables necessarily produces either three 0s and a 1 or three 1s and a 0.

Representation of Feedback Shift Registers

The behaviour of a feedback shift register has been described above by one or more of the following:

1. Its logical circuit.
2. A recurrence relation or logic function defining the feedback digit.
3. The characteristic polynomial consisting of powers of D.
4. A truth table giving the feedback digit in terms of the outputs of the binary stores.
5. The complete state diagram.
6. The sequences produced.

While each of these representations contains all the information about the system and the sequences it produces, certain properties are more readily manifested by one representation, others by another. For instance (1) gives no indication of the state diagram but does show if the system is linear and if the state diagram consists of separate cycles. On the other hand (5) indicates the length and number of sequences and readily enables them to be constructed, but does not directly reveal the form of the logic, and so on.

The following is a simple example showing how to arrive at the logic, given the complete state diagram shown in Fig. 9.3 (*a*). Since there are eight states the register contains three stores, and the general form is shown in Fig. 9.3 (*b*). The following truth table for y_r is derived from the state diagram:

y_r	y_{r-1}	y_{r-2}	y_{r-3}	Register State
1	0	0	0	0
0	0	0	1	1
0	0	1	0	2
0	0	1	1	3
1	1	0	0	4
0	1	0	1	5
1	1	1	0	6
0	1	1	1	7

The recurrence relation for y_r is therefore:

$$\begin{aligned} y_r &= \bar{y}_{r-1}\bar{y}_{r-2}\bar{y}_{r-3} + y_{r-1}\bar{y}_{r-2}\bar{y}_{r-3} + y_{r-1}y_{r-2}\bar{y}_{r-3} \\ &= \bar{y}_{r-3}(\bar{y}_{r-2} + y_{r-1}y_{r-2}) = \bar{y}_{r-3}(\bar{y}_{r-2} + y_{r-1}). \end{aligned}$$

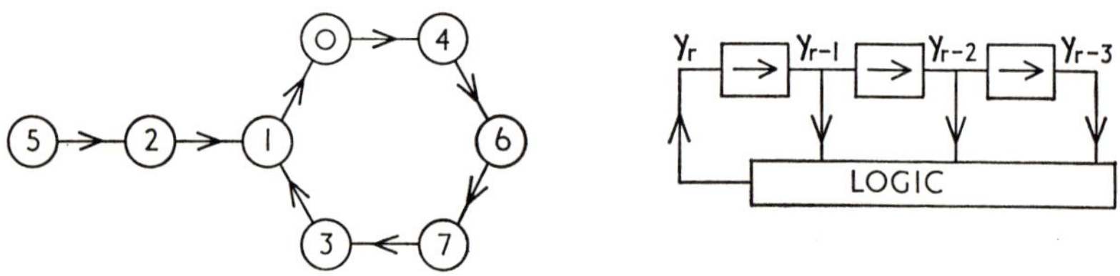

(*a*) (*b*)

Fig. 9.3. *Example of third-order state diagram and generating logic*

This is in its simplest form, and the equation contains the AND and OR correctives as well as inversions. It is always possible, however, to transform any logical equation or function into one containing the AND and ⊕ correctives and perhaps a logical one. Thus

$$\begin{aligned} y_r &= \bar{y}_{r-3}(\bar{y}_{r-2} + y_{r-1}) \\ &= \bar{y}_{r-3}[\overline{y_{r-2}\bar{y}_{r-1}}] \\ &= (y_{r-3} \oplus 1)[y_{r-2}(y_{r-1} \oplus 1) \oplus 1] \\ &= y_{r-1}y_{r-2}y_{r-3} \oplus y_{r-2}y_{r-3} \oplus y_{r-3} \oplus y_{r-1}y_{r-2} \oplus y_{r-2} \oplus 1. \end{aligned}$$

In polynomial form this equation becomes

$$(1 \oplus D \,.\, D^2 \,.\, D^3 \oplus D^2 \,.\, D^3 \oplus D^3 \oplus D \,.\, D^2 \oplus D^2)y_r = 1,$$

where the '1' inside the bracket denotes D^0. The characteristic polynomial is therefore

$$(1 + D \,.\, D^2 \,.\, D^3 \oplus D^2 \,.\, D^3 \oplus D^3 \oplus D \,.\, D^2 \oplus D^2)',$$

the bracket being primed to indicate the logical one in the equation, and this can conveniently be written in detached exponent notation as (0, 1 . 2 . 3, 2 . 3, 3, 1 . 2, 2)′.

The feedback function itself, namely the right-hand side of the equation for y_r, may be expressed as a sum of product terms π_i, where i is the decimal equivalent of the binary number formed by arranging the numerical part of the suffixes of each term in descending order. Thus:

$$y_{r-3}y_{r-2}y_{r-1} \equiv 111 \equiv \pi_7$$
$$y_{r-3}y_{r-2} \equiv 110 \equiv \pi_6$$
$$y_{r-3} \equiv 100 \equiv \pi_4$$
$$y_{r-2}y_{r-1} \equiv 11 \equiv \pi_3$$
$$y_{r-2} \equiv 10 \equiv \pi_2$$
$$1 \equiv 0 \equiv \pi_0$$

and so $$y_r = F(y_{r-1}, y_{r-2}, y_{r-3}) = \pi_0 \oplus \pi_2 \oplus \pi_3 \oplus \pi_4 \oplus \pi_6 \oplus \pi_7.$$

It is obvious that the terms of a linear feedback function, being isolated, correspond to the product terms of the form $\pi(2^k)$. The notation is described by Green and Dimond,[28] who show how, by expressing each minterm as a sum of product terms, a feedback function in polynomial form may readily be derived from the feedback function truth table, and vice versa. The same authors[29] also discuss the multiplication of functions in polynomial form and show that if X and Y are feedback shift registers, the sequences formed by the cascades $X \to Y \to$ output and $Y \to X \to$ output satisfy the same equation if and only if X and Y are both linear. Thus, for example, if X and Y are described by the equations $x_r = x_{r-1} \oplus x_{r-2}$ and $y_r = y_{r-1} \cdot y_{r-2}$, respectively, the overall feedback functions

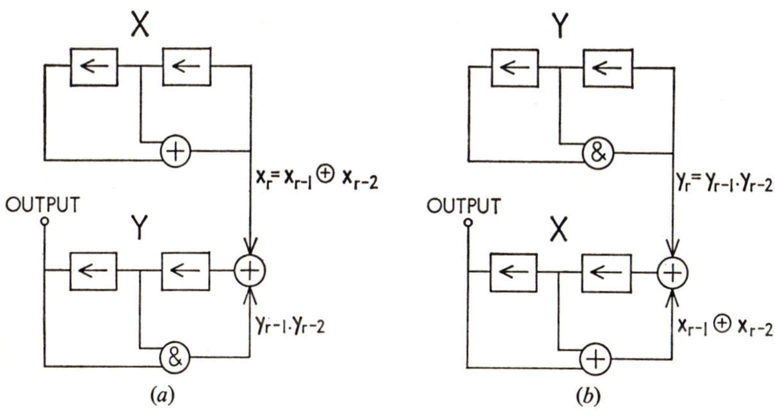

Fig. 9.4. *Non-commutative cascades of feedback shift registers*

of the arrangements shown in Figs. 9.4 (*a*) and (*b*) treated as single-shift registers are respectively (1, 2, 1. 2, 2 . 3, 3 . 4) and (1, 1 . 2, 3, 1 . 3, 1. 4, 2 . 4, 3 . 4) in detached exponent notation.

Whereas in chapter four it was shown how the cycle structure associated with any linear equation can be constructed, no corresponding procedure exists as yet for the non-linear case, nor even has an analysis been attempted of the cycle structure associated with an equation of the form $[F(D)]^n y_r = 0$, where $F(D)$ is a simple non-linear function.

Chapter ten

Applications of Binary Sequences

The applications of binary sequences are so numerous and so varied that a single chapter can only do scant justice to them or to the great ingenuity so many of them display. In general terms binary sequences find their main applications in the fields of radar, communications, and the evaluation of system parameters, but there are many other miscellaneous applications, a few of which will be mentioned here.

Binary Codes

It is appropriate at this stage to distinguish between sequences and codes, although the principles involved in their construction and use have much in common. A binary sequence, pseudo-random or otherwise, is normally a repetitive sequence of digits formed by some well-defined deterministic process, whereas a binary code is a dictionary of code *words*, each word consisting of a series of binary digits of finite length representing a particular piece of information. Codes are therefore associated with the problem of communication. Binary codes whose words are all of the same length are called Uniform Binary Codes.

The construction of *optimum codes* in given situations constitutes a highly sophisticated science to which an enormous amount of effort has been directed. Briefly, an optimum code is a set of transmitted signals chosen to make the likelihood of their correct interpretation at the receiver a maximum. Thus one requirement of an optimum code is maximum distinguishability between its components, i.e. the cross-correlation between any pair of code words is as small as possible. Evidently the time taken to attain correct interpretation is another essential parameter in assessing a code, and the information theory concepts of channel capacity, redundancy, and bandwidth are therefore indispensable to the theory of coding. For the noise-free channel the maximum rate of information is achieved using codes devised by Shannon, Fano, and Huffman. When it is feasible to suppose that channel noise has the effect of causing isolated random errors in the digit stream, so-called error-detecting and error-correcting codes are appropriate. In these a proportion of the digits of each code word is assigned to the detection and location of digit errors (anywhere in the code word). This is therefore an instance of the use of redundancy to overcome signal corruption due to noise. Codes of this kind were first proposed in 1950 by Hamming[31] and further studied by Reed, Muller, Golay, Slepian, Bose, Chaudhuri, and others.

The other situation in which codes are widely used is where the received signal, far from being only slightly degraded by noise, is in fact totally buried in noise. In this case it can be shown[32] that, for the case where the code words occur with equal probability and the noise added to the signal during transmission is white, the optimum code is again one whose components are as distinct as possible, and the best method of ascertaining which code word has been transmitted is to cross-correlate the received signal with each of the code words in turn. The one which gives the greatest correlation is the one most likely to have been transmitted. The codes used in this context are *orthogonal*, *trans-orthogonal*, and *bi-orthogonal*, and their performance is fully evaluated by Viterbi.[33]

The problem of achieving synchronization between transmitted and received signals has led to the construction of *comma-free codes*. A common-free code is one whose code words, which are again groups of binary digits usually of equal length N, are such that within any sequence of $2N - 1$ consecutive digits selected from a succession of code words only one code word is present. It is therefore possible to establish synchronization, and so enable a continuous message to be interpreted, by examining the first $2N - 1$ digits received. The advantage of such a code is that no special provision has to be made for synchronization, such as the periodic inclusion of a synchronizing pattern in the message sequence or the addition of a separate synchronizing channel. Comma-free codes are discussed by Stiffler,[18] Golomb,[34] Gilbert,[35] and others.

Coding theory is a very large subject in its own right and cannot be discussed further here.

Radar

A major task of radar is the measurement of the range of a target. This is usually accomplished by transmitting a pulse-modulated carrier and measuring the time delay between a transmitted pulse and the corresponding received pulse. If the maximum range to be measured is d, the interval between successive transmitted pulses must exceed $\frac{2d}{c}$ seconds to avoid ambiguity. On the other hand the range resolution is of the order of the pulse length $c\Delta$ in space, where Δ is the pulse duration. Thus, to measure very long ranges with great precision requires a very low duty cycle, in which energy is transmitted for only a minute proportion of the pulse recurrence period.

While a pulsed carrier is ideal for the simultaneous range estimation of a number of targets in the same direction, adequate performance at long ranges requires a very large pulse power. By using a carrier modulated by a pseudo-random sequence of clock period Δ and length $N = \frac{2d}{c\Delta}$ the same performance may be achieved for the same average transmitted power. In such systems *bi-phase modulation* is normally used, whereby the carrier amplitude is kept constant (thereby ensuring constant transmitted power) and the carrier phase is reversed whenever the level of the modulating sequence changes. The range may be measured directly by cross-correlating the sequence formed after demodulating the echo signal with a delayed version of the modulating sequence. When the delay equals that of the echo the two sequences are in phase and the cross-correlation is maximum. For any other delay the correlation is almost zero. A system of this kind has been used by Cabion[36] to measure the heights and reflecting properties of ionospheric layers. In this system the integration involved in cross-correlation is achieved very simply by means of a low pass filter.

A more usual requirement is the continuous estimation of the range of a moving object. In this case it is desirable to alter the delay automatically to maintain a state of maximum correlation. This is normally accomplished by means of a delay-lock tracker (for example, Gill[37]), in which an error voltage proportional to the difference between the echo and locally generated delays is used to alter the clock frequency of the modulating sequence so as to keep the sequence and the delayed version of the modulating sequence in phase.

An example of an *X*-band radar using pseudo-random modulation is described by Kramer,[38] and an interesting discussion of the effects of *incidental modulation*, i.e. spurious amplitude modulation and spurious departure from the 0^0 and 180^0 phase shifts associated with true bi-phase modulation, is given by Lichtenstein,[39] who also contrasts Kramer's system and an alternative system using a slightly different form of demodulation.

Pseudo-random modulation has also been used in Sonar detectors, and a paper by Stewart[40] compares the performance of four different sonar detecting systems using pseudo-random and other forms of modulation.

Perhaps the most elaborate radar using cross-correlation techniques is the one described by Easterling[41] for tracking the range of space vehicles. Here, owing to the ranges involved and the great range precision required, enormously long sequences have to be used, and it would be quite impractical to use a simple pseudo-random modulating sequence giving finite cross-correlation only when the received and delayed version of transmitted sequences are exactly in phase since the process of range acquisition might conceivably involve cross-correlating the echo sequence with every possible delayed version of the transmitted sequence in turn. Instead, the modulating sequence is formed by combining a number of short pseudo-random sequences of relatively prime periods $v_1, v_2, v_3, \ldots$ The resulting sequence, which has period $v_1v_2v_3 \ldots$, has a number of peaks in its autocorrelation function occurring at delays $v_1, v_2, v_3, \ldots, v_1v_2, v_1v_3, v_2v_3, \ldots, v_1v_2v_3, \ldots$, etc. The autocorrelation is different at the different peaks, being greatest when the delay equals the period of the combined sequence. Range acquisition is therefore effected in stages, the delay of one component in the comparison sequence being stepped till a peak in the cross-correlation between it and the echo sequence is reached, then that of the next component, till another (greater) peak is reached, and so on till maximum cross-correlation is attained. Thereafter a delay-lock system keeps track of the range.

As an example of a combined sequence with multiple peaks in its autocorrelation consider the sequence Z of length 105 formed by the termwise *multiplication* of the m-sequences $X = 1110100$, repeats, and $Y = 111100010011010$, repeats. In deriving the correlation between Z and a phase-shifted version of Z we termwise *add* them and compute the fraction: number of 0s $-$ number of 1s $\div$ number of 0s $+$ number of 1s over one period of Z for all possible phase shifts. Denote a component of the latter sequence by a capital letter if it is in phase with the corresponding component of the former sequence, and by a small letter if it is not. Thus XY is the former sequence, and either xY, Xy, xy, or XY is the latter. Since the lengths of the components are relatively prime, concepts of probability may be used to calculate the above fraction for all phase shifts.

Consider the correlation between sequences XY and xY. The following table gives the values of the termwise additions implied by the sum $XY \oplus xY$:

		Xx			
		00	01	11	10
Y	0	0	0	0	0
	1	0	1	0	1

The probability that Y is $0 = \frac{7}{15}$ and that it is $1 = \frac{8}{15}$. The probability that Xx is $00 = \frac{1}{7}$, that it is $01 = \frac{2}{7}$, that it is $11 = \frac{2}{7}$, and that it is $10 = \frac{2}{7}$.

Thus we can draw up a table indicating the probabilities of each cell in the above table as follows:

		Xx			
		1/7	2/7	2/7	2/7
Y	7/15	7/105	14/105	14/105	14/105
	8/15	8/105	16/105	16/105	16/105

The probability that the termwise sum is 0 is therefore $\frac{7}{105} + \frac{14}{105} + \frac{14}{105} + \frac{14}{105} + \frac{8}{105} + \frac{16}{105} = \frac{73}{105}$, and that it is 1 is $\frac{16}{105} + \frac{16}{105} = \frac{32}{105}$. The correlation is therefore $\frac{73 - 32}{105} = \frac{41}{105}$.

In a similar way we find that the correlation between XY and Xy is $\frac{41}{105}$. To calculate the correlation when both components are out of phase the corresponding tables are drawn up for the termwise addition $XY \oplus xy$ as follows:

		Yy			
		00	01	11	10
Xx	00	0	0	0	0
	01	0	1	1	0
	11	0	1	0	1
	10	0	0	1	1

and

		Yy			
		3/15	4/15	4/15	4/15
Xx	1/7	3/105	4/105	4/105	4/105
	2/7	6/105	8/105	8/105	8/105
	2/7	6/105	8/105	8/105	8/105
	2/7	6/105	8/105	8/105	8/105

from which the probability that the termwise sum is $0 = (3 + 4 + 4 + 4 + 6 + 8 + 6 + 8 + 6 + 8)/105 = \frac{57}{105}$ and that it is $1 = (8 + 8 + 8 + 8 + 8 + 8)/105 = \frac{48}{105}$. Thus the correlation is $\frac{57 - 48}{105} = \frac{9}{105}$.

When both components are in phase the correlation is 1, so there are three levels of correlation $\frac{9}{105}$, $\frac{41}{105}$, and 1, and the number of steps needed to bring the comparison sequence into phase with the echo sequence is at most $6 + 14 = 20$.

Communication

Most applications in this field make use of codes, which have been referred to above, but the following, which are not of this kind, should be mentioned. Two papers[42, 43] describe the use of pseudo-random sequences in multiplexing. The first considers a straightforward communication system in which a number of binary signals D_1, D_2, D_3, . . . are respectively multiplied by binary m-sequences S_1, S_2, S_3, . . . of equal length. The transmitted signal comprises the sum of these products. At the receiver the different signals are separated out by correlating the received signal with each of the sequences S_1, S_2, S_3 . . . Analysis shows that the cross-talk, i.e. the inference at the output of any one channel i, say, by signals $D_j(j \neq i)$, is determined primarily by the correlation between the sequences, being least when this is a minimum. Interestingly enough, the maximum out-of-phase cross-correlation between any two m-sequences of equal period is least when their period is prime, and becomes progressively greater as the number of prime factors in the period increases. Cross-correlation between sequences of equal period is briefly discussed in chapter eight.

The other paper describes a multichannel telegraph system using majority logic in the multiplexing process. Each of n binary (0, 1) data signals is added (modulo two) at the transmitter to one output (0, 1) of an n-stage shift register. By employing linear feedback over m stages (four, in the equipment described) the register outputs are successively phase-shifted versions of an m-sequence of period $2^m - 1$. The outputs of the modulo-two gates are fed to a multiplexer which is designed to produce a 1 if more than $n/2$ of its inputs are 1, and a -1 if less than $n/2$ inputs are 1. At the receiver the multiplexed signal is multiplied by the n outputs of a similar shift register arranged to run in synchronism with the one at the transmitter. Integration of each product signal then yields the corresponding original data signal. The theory of demultiplexing signals multiplexed by majority logic is explained by Titsworth.[44]

The basic principles of linear binary filters are described in chapter one, where it was shown that a binary signal could be recovered by passing it through a binary filter in the form of a feedback shift register, and then passing the resulting sequence through the inverse feed-forward filter. This procedure forms the basis of a mobile communication system for police use described by Thompson.[45] Multi-channel facilities are provided by the inclusion of an addressing arrangement, and the system affords a high degree of secrecy since the data signal is effectively scrambled by the filter in the transmitter and unscrambled by the inverse filter in the receiver. The information, normally in the form of speech, is converted into a sequence of binary digits by delta modulation, a delta modulator being a device which compares the current value of the speech signal with its value one clock period previously and produces a $+1$ or a -1 according to whether the former level is greater or less than the latter. The use of delta modulation avoids the need for transmitting synchronizing information.

Another important use of pseudo-random sequences of a very different nature is described by Ingram.[46] The standard method of assembling large numbers of telephony signals for transmission over wideband line or radio systems is the frequency-division multiplex technique. In this a large number of carriers of various related frequencies and equal amplitude are required, and the conventional methods of producing these are very expensive. The use of sequences having appropriate frequency spectra for providing these carriers is suggested. The paper first shows that a periodic pattern of pulses whose out-of-phase autocorrelation is low tends to have a uniform frequency spectrum, the extreme case being white noise, whose out-of-phase autocorrelation is zero and whose frequency spectrum is flat. The *m*-sequence is therefore ideal in this context. However, it is frequently desirable to generate spectra containing only specified harmonics, and the paper goes on to describe how sequences having the desired spectra may be produced by suitable logical combinations of given sequences. The basic logical connectives are modulo-two addition, AND and OR, and the essence of the analysis is that if corresponding terms A and B of two sequences consisting of 1s and −1s are combined using one or other of these connectives, the truth tables appropriate for spectral analysis are:

A	B	$A \oplus B = -AB$	A OR $B = \frac{1}{2}(A + B + 1 - AB)$	A AND $B = \frac{1}{2}(A + B - 1 + AB)$
−1	−1	−1	−1	−1
−1	+1	+1	+1	−1
+1	−1	+1	+1	−1
+1	+1	−1	+1	+1

Thus if the sequences are expressed as Fourier Series F_A and F_B, termwise addition (modulo-two) of the sequences, for example, produces a sequence whose Fourier series is minus the algebraic product of F_A and F_B, and similarly for the other connectives. Hence it is immediately possible to deduce the frequencies present in a combination of two sequences in terms of the frequencies present in the sequences themselves. The generation of a wide variety of spectra using combinations of two or more sequences having harmonically related clock frequencies is possible using this technique.

Other uses of binary sequences in the field of communications are mentioned, for example, by Noakes,[47] Roberts,[48] and Ramsden.[49] Another application, to the problem of character recognition, is described by Buykes,[50] and yet another use, as a perturbation signal in flow measurement, is mentioned by Abeysekera.[51]

Evaluation of System Parameters

Traditional methods of evaluating the parameters of linear systems fall into two categories, namely frequency domain and time-domain methods. In the former the system transfer function is measured directly at a number of frequencies by superimposing a sinusoidal disturbance on the input and comparing the amplitude and phase of input and output disturbances. This is time-consuming since a measurement at any one frequency can only be made after all transients due to the application of the sine wave have died away. Also in processes where noise is significant the measurements of amplitude and phase at each frequency must be averaged over a long time, and so run into the difficulty of holding the process steady during the measurements.

In the latter method the time response of the system to an input disturbance consisting of either an impulse or a step is measured. In the presence of noise, however, the disturbing signal has to be so large that the normal operation of the process may be affected. The alternative of repeating the tests a number of times and taking the average, only feasible anyway if the noise is fairly small, is again very time-consuming. In theory, however, both frequency and time-domain methods define the system since the transfer function is the Fourier transform of the impulse response, which is itself the derivative of the step response.

The shortcomings of the above methods are largely overcome by using a correlation technique. A disturbing signal in the form of a pseudo-random binary sequence is added to the normal system input, and the resulting output disturbance is cross-correlated with the input. If the sequence parameters are suitable, and subject to certain approximations, the cross-correlation, as a function of time, is directly proportional to the impulse response of the system. The cross-correlation is generally carried out by a digital computer, though it is worth remarking that since the input disturbance is a binary signal the multiplication involved in cross-correlation may readily be accomplished using a relay (electronic or otherwise). Furthermore, since the computation $\phi_{xy}(\tau) = \underset{T\to\infty}{\mathrm{Lt}}\frac{1}{2T}\int_{-T}^{+T} x(t)y(t+\tau)dt$ amounts to finding the mean value of the product $x(t)y(t+\tau)$ it can be carried out using a low pass filter. Since noise in the output is uncorrelated with the input signal its contribution to the cross-correlation is small, and becomes smaller the longer the correlation time. Sensible results, in fact, can be obtained even with very low signal-to-noise ratios provided that the process can be held steady over a long time.

The correlation technique was first used in 1961 by Balcomb[52] to investigate the parameters of a nuclear rocket engine, and has since been successfully applied to a large number of full-scale plants and processes such as nuclear reactors, chemical plants, steel furnaces, rolling mills, oil refineries, paper mills, industrial boilers, and evaporators, etc. A very extensive list of applications is included in an excellent paper on the subject by Godfrey.[53]

Theory of Correlation Technique

This method is now so important that it will be explained in some detail.

Suppose a noise-free system operating at some stationary level. Let a signal $x(t)$ be applied as a disturbance to the standing value of the input, and let $y(t)$ be the corresponding disturbance at the output. Assume that these disturbances are sufficiently small for the system to behave linearly. In this case the Fourier transforms $X(w)$ and $Y(w)$ of $x(t)$ and $y(t)$ are related by the equation $Y(w) = A(w)X(w)$, where $A(w)$ is the system transfer function (at this particular operating level).

If $x(t)$ is a unit impulse applied at $t = 0$, $y(t)$ is the impulse response of the system, usually denoted by $h(t)$. Since $X(w)$ for such an input is unity, the transform of $h(t)$ is the system transfer function. The above equation may therefore be written $Y(w) = H(w)X(w)$, where $H(w)$ is the transform of $h(t)$.

The Convolution Theorem (see, for example, Helstrom[54]) states that if time functions $f_1(t)$ and $f_2(t)$ have Fourier transforms $F_1(w)$ and $F_2(w)$, the product $F_1(w)F_2(w)$ is the transform of the *convolution* of $f_1(t)$ and $f_2(t)$. The convolution of $f_1(t)$ and $f_2(t)$ is usually denoted by $f_1(t)*f_2(t)$ and equals $\int_{-\infty}^{+\infty} f_1(s)f_2(t-s)ds$. Applying this theorem to the above system, $Y(w)$ is the transform of $h(t)*x(t)$, and so $y(t)$ itself equals $\int_0^{Ts} h(s)x(t-s)ds$, since the impulse response $h(t)$ is zero for $t < 0$ and is again zero after some time T_s, known as the settling time of the system. Given a suitable form of $x(t)$ this formula enables $h(s)$ to be obtained from measurements of $x(t)$ and $y(t)$. The time for which these have to be made to be able to evaluate $h(s)$ depends on the nature of the waveform $x(t)$. In particular if $x(t)$ is a unit DC level $y(t)$ equals $\int_0^{Ts} h(s)ds$, the DC gain of the system, and $h(s)$ itself can never be found. On the other hand if $x(t)$ is an impulse $K\delta(t)$, $y(t) = Kh(t)$,

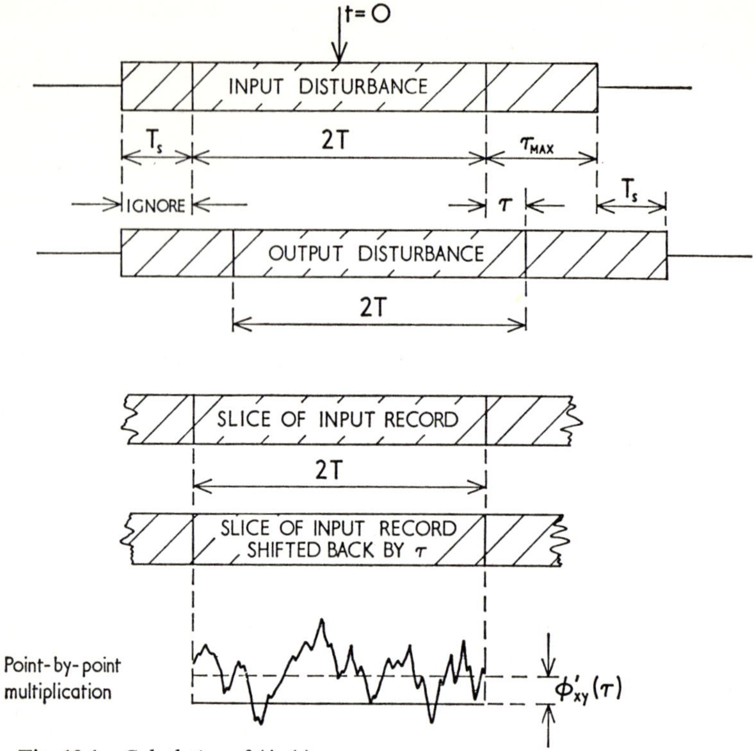

Fig. 10.1 *Calculation of* $\phi'_{xy}(\tau)$

and the whole impulse response can be found in time T_s. In this case if the output of the system is contaminated by noise $n(t)$ uncorrelated with the input or output, the contaminated output $z(t) = y(t) + n(t)$, and $h(t)$ is then given by the formula $z(t) = Kh(t) + n(t)$; $h(t)$ may then be found by averaging a number of measurements of $z(t)$.

In the correlation method an input disturbance $x(t)$ is applied for a certain length of time, and $x(t)$, and the corresponding output disturbance $y(t)$, are recorded. The cross-correlation $\phi'_{xy}(\tau)$ is then computed for various values of τ according to the formula $\phi'_{xy}(\tau) = \frac{1}{2T}\int_{-T}^{+T} x(t)y(t+\tau)dt$, and the autocorrelation $\phi'_{xx}(\tau - s)$ of the input disturbance is likewise computed for the same values of τ and various values of s according to the formula $\phi'_{xx}(\tau - s) = \frac{1}{2T}\int_{-T}^{+T} x(t)x(t+\tau-s)dt$. The correlations are primed since T is not taken to infinity. To obtain $h(s)$ from $\phi'_{xy}(\tau)$ and $\phi'_{xx}(\tau - s)$ we observe that:

$$\phi'_{xy}(\tau) = \frac{1}{2T}\int_{-T}^{+T} x(y)y(t+\tau)dt$$

$$= \frac{1}{2T}\int_{-T}^{+T} x(t)\left\{\int_{0}^{T_s} h(s)x(t+\tau-s)ds\right\}dt$$

$$= \int_{0}^{T_s} h(s)\left\{\frac{1}{2T}\int_{-T}^{+T} x(t)x(t+\tau-s)dt\right\}ds$$

$$= \int_{0}^{T_s} h(s)\phi'_{xx}(\tau-s)ds$$

Figs. 10.1 and 10.2 show how $\phi'_{xy}(\tau)$ and $\phi'_{xx}(\tau - s)$ are computed from the recordings of input and output disturbances.

The formula for $\phi'_{xy}(\tau)$ shows that the required range for s is $0 \leqslant s < T_s$. By comparing it with the corresponding formula for $y(t)$ we see that the range of values $0 \leqslant \tau < \tau_{\max}$ of τ needed to enable $h(s)$ to be evaluated

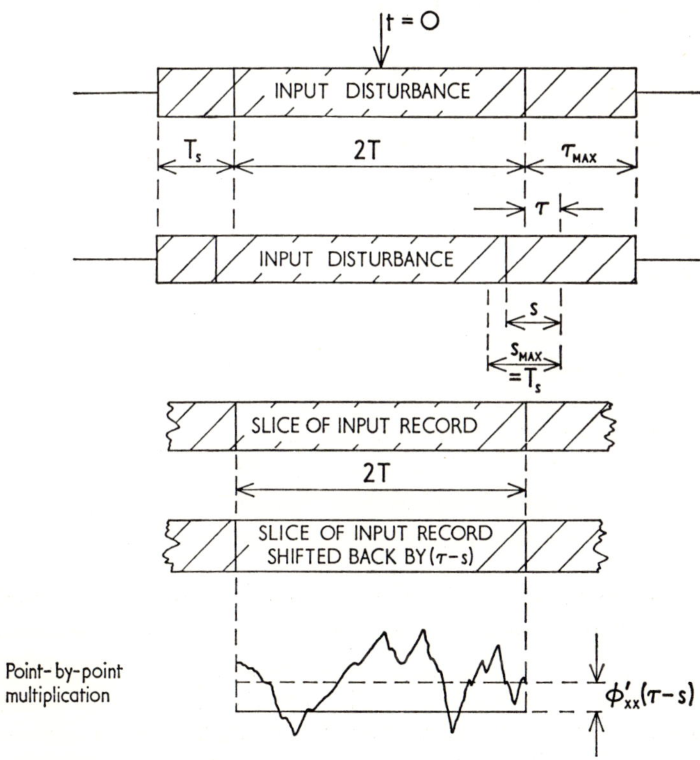

Fig. 10.2. *Calculation of* $\phi'_{xx}(\tau - s)$

depends on the form of $\phi'_{xx}(\tau)$. For an input disturbance consisting of white noise of power spectral density P_0, $\phi_{xx}(\tau)$ is $P_0\delta(\tau)$. For T large enough $\phi'_{xx}(\tau)$ becomes equal to $\phi_{xx}(\tau)$, and in this case $\phi'_{xy}(\tau) = \int_0^{T_s} h(s)P_0\delta(\tau - s)ds$ $= P_0 h(\tau)$. The impulse response is therefore the computed value of $\phi'_{xy}(\tau)$ divided by P_0, and as in the corresponding case above $\tau_{\max}$ is T_s, the duration of the impulse response.

Periodic Input Disturbance

The trouble with white noise is that to make $\phi'_{xx}(\tau)$ equal to $\phi_{xx}(\tau)$, T has to be very large, and for this reason an input disturbance in the form of periodic noise is used instead. In this case we have:

$$\phi_{xx}(\tau) = \underset{T\to\infty}{\text{Lt}} \frac{1}{2T}\int_{-T}^{+T} x(t)x(t+\tau)dt$$

$$= \frac{1}{T_0}\int_0^{T_0} x(t)x(t+\tau)dt,$$

where T_0 is the period of $x(t)$. Also, since $x(t-s) = x(t+nT_0-s)$, $n = \pm 1, \pm 2$, etc.,

$$y(t+nT_0) = \int_0^{T_s} h(s)x(t+nT_0-s)ds$$

$$= \int_0^{T_s} h(s)x(t-s)ds$$

$$= y(t),$$

so $y(t)$ is likewise periodic with period T_0. Therefore

$$\phi_{xy}(\tau) = \underset{T\to\infty}{\text{Lt}} \frac{1}{2T}\int_{-T}^{+T} x(t)y(t+\tau)dt = \frac{1}{T_0}\int_0^{T_0} x(t)y(t+\tau)dt.$$

The somewhat indeterminate correlation time $2T$ may, indeed must, now be made an exact multiple of T_0. The impulse response is thus derived from the formula $\phi_{xy}(\tau) = \int_0^{T_s} h(s)\phi_{xx}(\tau)ds$, where ϕ_{xy} and ϕ_{xx} may be computed *exactly* from *finite* records of $x(t)$ and $y(t)$.

The form of periodic noise most commonly used is the pseudo-random binary sequence. If this has length v, binary levels $\pm a$, and clock period δ, its autocorrelation function is as shown in Fig. 10.3.

The terms of the integrand $h(s)\phi_{xx}(\tau - s)$ for this case are shown in

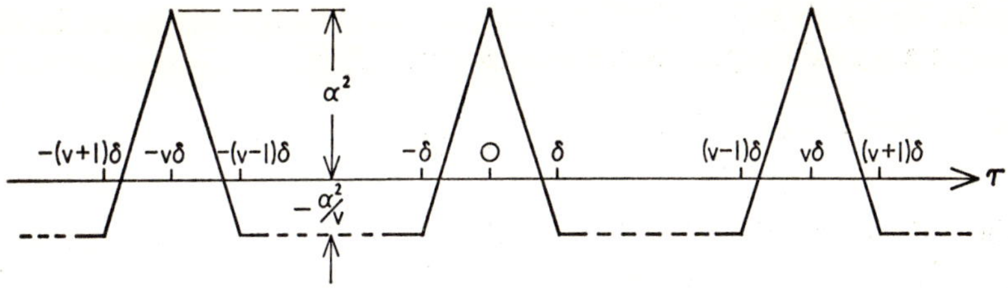

Fig. 10.3. *Autocorrelation function of a pseudo-random sequence*

Fig. 10.4. It is apparent that provided $v\delta > T_s$, and 2δ is small enough to allow the assumption that $h(s)$ is sensibly constant during the autocorrelation spike, $\phi_{xx}(\tau - s)$ may be regarded, at least down to $\tau = \delta$, as a single impulse of area $a^2\left(1 + \frac{1}{v}\right)\delta$ occurring at $s = \tau$, plus a DC level $-\frac{a^2}{v}$. Consequently $\phi_{xy}(\tau) = a^2\left(1 + \frac{1}{v}\right)\delta \, . \, h(\tau) - \frac{a^2}{v}\int_0^{T_s} h(s)ds$. As shown above, the integral is the DC gain of the system, and after allowing for this $h(\tau)$ follows from a knowledge of the sequence parameters and the computed value of $\phi_{xy}(\tau)$. Fig. 10.5 shows the records of $x(t)$ and $y(t)$ required for the evaluation of the impulse response in this case.

System Noise

Noise in the system output has been mentioned above in connection with the direct measurement of impulse response. In the correlation method we are concerned with its contribution to the input-output cross-correlation rather than to the output itself. If the contaminated output is given as

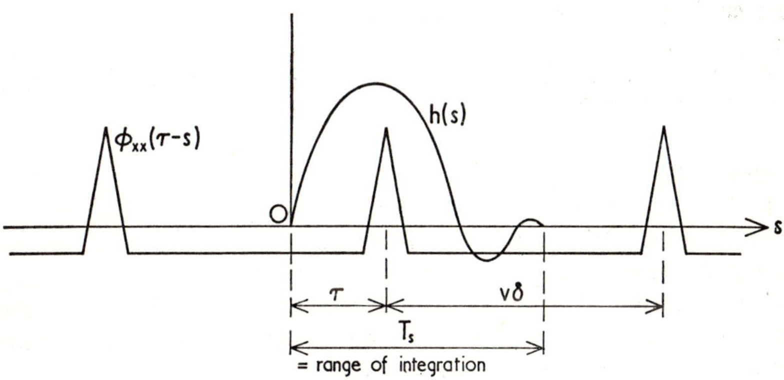

Fig. 10.4. *Terms of the integrand* $[h(s)\phi_{xx}(\tau - s)]$

before by $z(t) = y(t) + n(t)$ the measured input-output correlation in the case of a periodic input disturbance of period T_0 is

$$\phi_{xz}(\tau) = \frac{1}{NT_0}\int_0^{NT_0} x(t)z(t+\tau)dt = \phi_{xy}(\tau) + \phi'_{xn}(\tau)$$

where the correlation time is a multiple N of the period T_0. The term $\phi_{xy}(\tau)$ is unaffected by N since ϕ_{xy} has period T_0. However, since noise is incoherent, any summation of noise signals must be effected by adding noise powers. The summation of N noise signals of equal RMS value therefore yields a noise signal whose RMS value is only $\sqrt{N}$ times greater. Thus $\phi'_{xn}(\tau)$ is inversely proportional to $\sqrt{N}$, and therefore diminishes as N

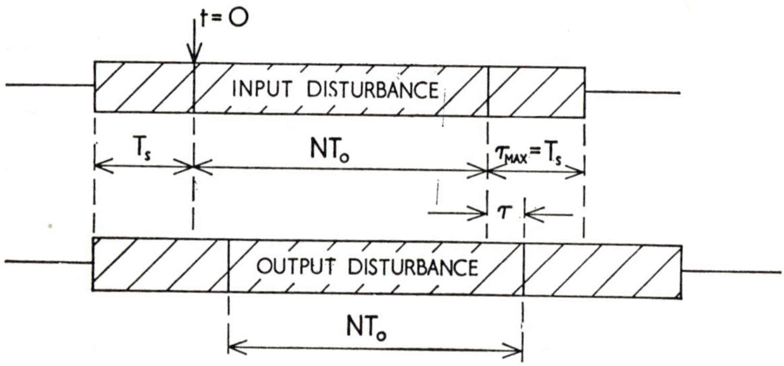

Fig. 10.5. *Records of $x(t)$ and $y(t)$ for periodic noise input*

is increased. We also observe that if $x(t)$ is a pseudo-random sequence having levels $\pm a$, $\phi_{xy}(\tau)$ is proportional to a^2, while $\phi'_{xn}(\tau)$ is proportional to a, so the greater the sequence amplitude the smaller the effect of noise, as expected.

An analysis of a noisy system, together with numerous references, is given by Macleod.[55]

System Drift

In the course of an experiment to determine a system's impulse response the system may drift. Drift may be regarded as a slow time-dependent output $d(t)$ superimposed on the output disturbance $y(t)$ in the same way as is system noise. Its effect is to introduce what appears to be noise on the cross-correlation function, and one way of reducing it is to pass the output signal through a high-pass filter.

The usual analysis of the effect of drift begins by assuming $d(t)$ to be of the form $a_0 + a_1t + a_2t^2 + \ldots$ For $x(t)$ having period T_0 and correlation over N periods, $\phi_{xz}(\tau) = \phi_{xy}(\tau) + \phi_{xd}(\tau)$, where $\phi_{xd}(\tau)$ is the sum of terms depending on the *time-moments* $\int_0^{NT_0} t^i x(t)dt$ of the input disturbance $x(t)$. Using a pseudo-random binary sequence as input it is possible to minimize these terms by starting the cross-correlation at a suitable point in the sequence. Alternatively a more general approach is to use a weighted input $w(t)x(t)$, where $w(t)$ is a weighting function chosen to eliminate the contribution $\phi_{xd}(\tau)$ altogether. Another very recent approach starts off by expressing the drift as a Fourier series. A great deal of attention has been paid to the subject of drift, and ref. 56 lists the relevant contributions.

System non-linearities and the relevance of inverse-repeat sequences in this connection are discussed in ref. 53 and ref. 57, while errors in cross-correlation measurements due to imperfect, i.e. non-rectangular, binary sequence inputs are discussed in ref. 58.

There is no doubt that the correlation technique for measuring the dynamics of full-scale processes under normal operating conditions can now be considered as conventional, though applications so far reported are incomplete in two respects. Firstly, they have all treated the identification problem as separate from the control problem. No attempt has yet been made to combine identification with on-line control. Secondly the applications have assumed a linear system, or at least one which behaves linearly over finite ranges of input. Recent work using p-nary rather than binary sequences as stimulating signals of non-linear systems may well prove significant in this context. One area still unexploited is the identification of *multivariable systems*, for which pseudo-random signals, suitably modified, can separate out the effect which each input channel has on each output channel from one experiment.

Conclusion

In this chapter an attempt has been made to review the more important and representative applications of binary sequences. Most of these depend primarily on the correlation properties of sequences, and it seems likely that these will dominate future applications. Despite the straightforward basic theory of linear sequences new properties are still coming to light, and with the growing interest in multi-level sequences and their possible applications, this will surely continue to occur. Theoretical investigations into the mathematical properties of non-linear shift register sequences, however, though interesting, have not as yet suggested new applications based on any other than correlation properties.

Appendix

Moments of Binary Sequences having Multipliers

In chapter ten it was pointed out that when evaluating a system's impulse response using the correlation technique system drift occurring during the experiment introduces an error which depends on the 'moments' of the test signal. The scheme proposed by H. A. Barker (ref. 56 (vi)) for minimizing this error is based on the fact that, for any *m*-sequence of length $2^n - 1$, a digit in the sequence can always be found for which the first moment $\sum_{r=1}^{2^n-1} r \cdot x_r$ computed from this digit is 2^{n-1}, where x_r is ± 1 according as the *r*th digit of the sequence is 1 or 0. Ream (ref. 56 (ix)) has proved this, and in this appendix the moments of binary sequences having multipliers is considered generally.

A sequence $\{x_r\}$ has a multiplier *m* if the sequence formed by sampling every *m*th term of $\{x_r\}$ is a phase-shifted version of $\{x_r\}$. As shown in chapter seven, a term can always be found from which to begin sampling so as to make the two sequences termwise identical. Calling this term x_0 we therefore have equations $x_{\mu 1} = x_1, x_{\mu_2} = x_2, \ldots, x_{\mu_v} = x_v \, (= x_0)$, where v is the sequence period and $\mu_t = tm \pmod{v}, t = 1, 2, \ldots, v$.

For the class of sequences having multiplier 2 and period $2^n - 1$ (*m*-sequences belong to this class) these equations become $x_1 = x_2 = x_4 = \ldots, x_3 = x_6 = \ldots, x_5 = x_{10} \ldots$, etc. Taking the specific case of $n = 4$, $x_1 = x_2 = x_4 = x_8$, $x_3 = x_6 = x_{12} = x_9$, $x_5 = x_{10}$, and $x_7 = x_{14} = x_{13} = x_{11}$. The suffixes may be grouped into cosets as follows:

1	2	4	8
3	6	12	9
5	10		
7	14	13	11
15			

The last entry corresponds to the remaining term x_{15}. Each term represents the magnitude of the contribution of the corresponding sequence term to the first moment of the sequence. Thus if the sequence terms $x_{1,2,4,8}$ are each $+1$, the coset 1, 2, 4, 8 contributes $1 + 2 + 4 + 8 = 15$ to the moment, and if the terms $x_{7,14,13,11}$ are each -1, the corresponding coset contributes $-7 - 14 - 13 - 11 = -45$ to the first moment. It therefore remains to evaluate the sum of the terms of each coset in the general case.

If y is a term in a coset, the next term is $2y$ if $y < y/2$ and $2y - v$ if $y > v/2$. A coset can therefore be divided into groups of terms as shown below, each group containing one term greater than $v/2$, this being the last term of the group.

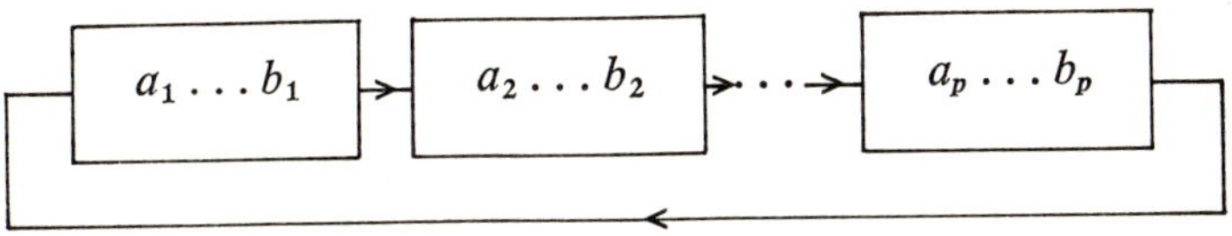

Since the terms of a group form a GP of common ratio 2, if a is the first term of a group and Σ is the sum of its terms, the first term of the next group is $\Sigma + a - v$. We have therefore:

$$\begin{aligned} a_2 &= \Sigma_1 + a_1 - v \\ a_3 &= \Sigma_2 + a_2 - v \\ &\ldots\ldots\ldots\ldots \\ a_p &= \Sigma_{p-1} + a_{p-1} - v \\ a_1 &= \Sigma_p + a_p - v \end{aligned}$$

Adding these equations gives $\Sigma_1 + \Sigma_2 + \ldots + \Sigma_p =$ sum of terms of a coset $= kv$, where k is the number of terms in the coset greater than $v/2$.

Let one period of the sequence consist of a 1s and b -1s. If the first moment, calculated over one period from a given term, is M, the moment computed over one period from the previous term is $M + [(a - b) - v]$ if the latter term is $+1$, and $M + [(a - b) + v]$ if it is -1. Thus each backward phase shift increments the moment by $(a - b) \pm v$. Let the value of the sequence terms corresponding to the sth coset be $\beta(s)$. The first moment of the sequence $x_1, x_2, \ldots, x_v$ is therefore $v\sum_s \beta(s)k(s)$, where $k(s)$ is the number of terms in the sth coset greater than $v/2$. Consequently the first moment of the phase-retarded sequence

$$x_{\left(\frac{v+1}{2}\right)}, x_{\left(\frac{v+3}{2}\right)}, \ldots, x_v, x_1, \ldots, x_{\left(\frac{v-1}{2}\right)} \text{ is}$$

$$\begin{aligned} v\sum_s \beta(s)k(s) + \left(\frac{v+1}{2}\right)(a - b) - v\sum_s \beta(s)k(s) &= \left(\frac{v+1}{2}\right)(a - b) \\ &= 2^{n-1}(a - b). \end{aligned}$$

For m-sequences, $a - b = 1$, which proves the result.

In the general case m and v are any numbers (in practice they have to be co-prime to lead to cosets). In particular if v is prime, non-trivial sequences only result if m is a quadratic residue of v, since only in this case is m, raised to a power less than v, congruent to 1 (mod v)), cosets may be formed as above. Each of these can be divided into groups of terms ending in a term lying between v/m and v, the previous terms lying between 0 and v/m. Since the terms of a group form a GP of common ratio m, if a is the first term of a group and Σ is the sum of its terms, the last term of the group is $((m-1)\Sigma + a)/m$. The first term of the next group is therefore $(m-1)\Sigma + a - \omega v$, where $\omega v < (m-1)\Sigma + a < (\omega + 1)v$, i.e. where $\omega(v/m) <$ last term of the group $< (\omega+1)(v/m)$. Thus ω is 1, 2, . . ., $(m-1)$ according as the last term of the group lies in the interval $(v/m, 2v/m)$, $(2v/m, 3v/m)$, . . ., $((m-1)v/m, v)$. As before,

$$
\begin{aligned}
a_2 &= (m-1)\Sigma_1 + a_1 - \omega_1 v \\
a_3 &= (m-1)\Sigma_2 + a_2 - \omega_2 v \\
&\cdots\cdots\cdots\cdots\cdots \\
a_p &= (m-1)\Sigma_{p-1} + a_{p-1} - \omega_{p-1} v \\
a_1 &= (m-1)\Sigma_p + a_p - \omega_p v
\end{aligned}
$$

By addition, $\Sigma_1 + \Sigma_2 + \ldots + \Sigma_p = (\omega_1 + \omega_2 + \ldots + \omega_p)v/(m-1)$, that is, the sum of the terms of a coset is the product of $v/m - 1$ and the total number of terms in the coset lying between v/m and $2v/m$ plus twice the total number of terms in the coset lying between $2v/m$ and $3v/m$ plus . . . plus $(m-1)$ times the total number of terms in the coset lying between $(m-1)v/m$ and v.

For example, with $v = 19$ and $m = 4$ the cosets, with terms in ascending order and separated into the above intervals, are:

1 4	5 6 7 9	11	16 17
2 3	8	10 12 13 14	15 18
19			

The sum of the terms of the first coset, which equals the magnitude of the contribution of the corresponding sequence terms to the first moment, is:

$$\frac{19}{3}(4 \times 1 + 1 \times 2 + 2 \times 3) = 76$$

The sum of the terms of the second coset is $\dfrac{19}{3}(1 \times 1 + 4 \times 2 + 2 \times 3)$ $= 95$.

Sequences defined by multipliers greater than two do not, of course, possess the symmetry necessary to provide a starting-point for which the first moment is independent of the distribution of 1s and -1s in the sequence.

Bibliography

1 ELSPAS, B.: *Autonomous Linear Sequential Networks*, Linear Sequential Switching Circuits, Ed. W. H. Kautz, Holden-Day, 1965, Ch. 2.

2 GODFREY, K. R., EVERETT, D. and BRYANT, P. E.: *Input Transducer Errors in Binary Cross-Correlation Experiments-2*, Proc. I.E.E., Jan. 1966, p. 185.

3 SCHOLEFIELD, P. H. R.: *Shift Register Generating* m-*sequences*, Electronic Technology, Oct. 1960.

4 PETERSON, W. W.: *Error-correcting Codes*, Wiley, 1961, Ch. 7.

5 GOLOMB, S.: *Shift Register Sequences*, Holden-Day, 1967, Ch. 3.

6 VITERBI, A. J.: *Power Spectra and Autocorrelation Functions*, Digital Communications with Space Applications, Ed. S. Golomb, Prentice-Hall, 1964, Ch. 3.

7 EASTERLING, M. F.: *Modulation by Pseudo-random Sequences*, Digital Communications with Space Applications, Ed. S. Golomb, Prentice-Hall, 1964, Ch. 5.

8 DAVIES, A. C.: *Probability Distribution of Pseudo-random Waveforms obtained from* m-*sequences*, Electronics Letters, March 1967.

9 DAVIES, A. C.: *Probability Distribution of Noiselike Waveforms Generated by a Digital Technique*, Electronics Letters, 20 Sept. 1968.

10 EVERETT, D.: *Periodic Digital Sequences with Pseudo-random properties*, G.E.C. Journal, Vol. 33, No. 3, 1966, p. 115.

11 HALL, J. Jr.: *A Survey of Difference Sets*, Proc. Amer. Math. Soc., Vol. 7, 1956, p. 975.

12 LEHMER, E.: *On Residue Difference Sets*, Canad. J. Maths., Vol. 5, 1953, p. 425.

13 STANTON, R. G. and SPROTT, D. A.: *A Family of Difference Sets*, Canad. J. Maths., Vol. 10, 1958, p. 73.

14 ZIERLER, N.: *Linear Recurring Sequences*, J. Soc. Ind. App. Maths., Vol. 7, Pt. 1, March 1959, p. 31.

15 CARMICHAEL, R. D.: *Groups of Finite Order*, Athaneum Press, 1937.

16 CAMPBELL, L. L.: *Two Properties of Pseudo-random Sequences*, Trans. I.R.E., March 1959, p. 32.

17 BARKER, R. H.: *Group Synchronizing of Binary Digital Systems*, Communication Theory, Ed. W. D. Jackson, Academic Press, 1953, p. 273.

18 STIFFLER, J. J.: *Synchronization Techniques*, Digital Communications with Space Applications, Ed. S. Golomb, Prentice-Hall, 1964, Ch. 8.

19 SIMPSON, H. R.: *Statistical Properties of a Class of Pseudo-random Sequences*, Proc. I.E.E., Dec. 1966, p. 2075.

20 BROWN, R. F.: *Drift Correction in Periodic Cross-correlation Schemes*, Electronics Letters, 1 Nov. 1968, p. 478.

21 BROWN, R. F. and GOODWIN, G. C.: *A New Class of Pseudo-random Binary Sequences*, Electronics Letters, May 1967, p. 198.

22 GOODWIN, G. C. and BELL, R. D.: *Error-free Identification using Maximally Orthogonal P.R.B.N. Sequences*, Electronics Letters, 1 Sept. 1969.

23 DARNELL, M.: *Generation of Quadratic Residue Sequences*, Electronics Letters, Feb. 1967, p. 88.

24 MARTIN, M. H.: *A Problem in Arrangements*, Bull. Amer. Math. Soc., Vol. 40, Dec. 1934.

25 DE BRUJN, N. G.: *A Combinational Problem*, Proc. Koninklijke Nederlandsche Akademie van Wetenschappen, Vol. 49, 1946, p. 758.

26 GOLOMB, S.: *Shift Register Sequences*, Holden-Day, 1967, Ch. 6.

27 BRYANT, P. R. and EVERETT, D.: *Theory of Feedback Shift Registers with Invertors*, Proc. I.E.E., Sept. 1969.

28 GREEN, D. H. and DIMOND, K. R.: *Polynomial Representation of Nonlinear Feedback Shift Registers*, Proc. I.E.E., Jan. 1970.

29 GREEN, D. H. and DIMOND, K. R.: *Nonlinear Product Feedback Shift Registers*, Proc. I.E.E., April 1970.

30 DOUCE, J. L.: *Barker Sequences*, Electronics Letters, April 1966, p. 159.

31 HAMMING, R. W.: *Error Detecting and Error Correcting Codes*, Bell Syst. Tech. J., April 1950, p. 147.

32 GOLDMAN, S.: *Information Theory*, Prentice-Hall, 1953, p. 230.

33 VITERBI, A. J.: *Phase-coherent Communication over the Continuous Gaussian Channel*, Digital Communications with Space Applications, Ed. S. Golomb, Prentice-Hall, 1964, Ch. 7.

34 GOLOMB, S. and GORDON, B.: *Codes with Bounded Synchronization Delay*, Information and Control, Vol. 8, 1965, p. 355.

35 GILBERT, E. N.: *Synchronization of Binary Messages*, Trans. I.R.E., IT-6, Sept. 1960, p. 470.

36 CABION, P. J. and VOLCKMAN, P. V.: *Pseudo-noise Ionospheric-Communication System*, Electronics Letters, 26 June 1969, p. 299.

37 GILL, W. J.: *A Comparison of Binary Delay-lock Tracking Loop Implementations*, I.E.E.E. Trans. AES-2, 1966, p. 415.

38 KRAMER, A., KELLEY, D., SOLOMON, A., BERKOVITS, G. and SCHERER, F.: *An Integrated X-band CW Radar Front End*, Microwave J., Vol. 11, 1968, p. 55.

39 LICHTENSTEIN, N. and RAUCH, S.: *Effect of Incidental Modulation in Systems Employing Pseudo-random Codes*, Electronics Letters, 9 Jan. 1969, p. 4.

40 STEWART, J. L. and WESTERFIELD, E. C.: *A Theory of Active Sonar Detectors*, Proc. I.R.E., 1959, p. 872.

41 EASTERLING, M. F.: *Applications to Ranging*, Digital Communications with Space Applications, Ed. S. Golomb, Prentice-Hall, 1964, Ch. 6.

42 JUDGE, W. J.: *Multiplexing using Quasi-orthogonal Binary Functions*, Trans. A.I.E.E. Pt. I (Communication and Electronics), May 1962, p. 81.

43 BARRETT, R. and KARRAN, J.: *Correlation-multiplex Data-transmission System*, Electronics Letters, 29 Nov. 1968, p. 538.

44 TITSWORTH, R. C.: *A Boolean-function Multiplexed Telemetry System*, I.E.E.E. Trans. SET-9, 1963, p. 42.

45 MILLER, L. S. and THOMPSON, W.: *Use of Pseudo-noise Concepts in Mobile Communications*, I.E.E. Conf. Publ. 64, Signal processing methods for radio telephony, May 1970.

46 INGRAM, D. G. W., WELLS, P., BRYANT, P. R. and EVERETT, D.: *Digital Techniques in Carrier Frequency Generation*, Proc. I.E.E., Feb. 1966, p. 243.

47 NOAKES, P. W.: *A Method of Analysing Errors Resulting from Digital Transmission over Telephone Lines*, I.E.R.E. Conf. on Digital Methods of Measurement, July 1969.

48 ROBERTS, L. G.: *Picture Coding using Pseudo-random Noise*, Trans. I.R.E. on Information Theory, Feb. 1962. p. 145.

49 RAMSDEN, A. N.: *The Measurement of the Performance Parameters of P.C.M. Links*, I.E.R.E. Conf. on Digital Methods of Measurement, July 1969.

50 BUYKES, J., MAN, A., STEIN, A., THOMPSON, L. and YOUSSEF, T.: *Stochastic Sampling Character-Recognition System*, Electronics Letters, 11 Dec. 1969, p. 635.

51 ABEYSEKERA, S. and BECK, M. S.: *Cross-correlation Techniques applied to Pulsating Flow Measurement*, Measurement and Control, June 1970, p. T109.

52 BALCOMB, J. D., DERNUTH, H. B. and GYFTOPOULOS, E. P.: *A Cross-correlation Method for Measuring the Impulse Response of Reactor Systems*, Nucl. Sci. Eng., 1961, No. 11, p. 159.

53 GODFREY, K. R.: *The theory of the Correlation Method of Dynamic Analysis and its Application to Industrial Processes and Nuclear Power Plant*, Measurement and Control, May 1969, p. T65.

54 HELSTROM, C. W.: *Statistical Theory of Signal Detection*, Pergamon, 1969.

55 MACLEOD, C. J.: *Parameter Estimation using Pseudo-random Binary Sequences*, Electronics Letters, 23 Jan. 1969, p. 35.

56 Proc. I.E.E.,

Dec. 1965, p. 2385.
Oct. 1967, p. 1582.

Electronics Letters,

July 1966, p. 243
Sept. 1966, p. 327
Feb. 1967, p. 72
Nov. 1967, p. 524
31 May 1968, p. 220
14 June 1968, p. 255
6 Sept. 1968, p. 380
1 Nov. 1968, p. 478.
27 Dec. 1968, p. 588
9 Jan. 1969, p. 15
20 March 1969, p. 112
1 May 1969, p. 180

57 Electronics Letters,

May 1966, p. 164
Dec. 1966, p. 453
April 1967, p. 151
31 May 1968, p. 224
13 Nov. 1969, p. 594
30 April 1970, p. 269

58 GODFREY, K. R., MURGATROYD, W., EVERETT, D. and BRYANT, P. R.: *Input-transducer Errors in Binary Cross-correlation Experiments —1, 2 and 3*, Proc. I.E.E., March 1965, p. 565, Jan. 1966, p. 185, and June 1966, p. 1095.

Index

AND function, vii
Antisymmetric Sequence, 83
Autocorrelation, xvi, 52, 59
Autonomous System, 3

Balanced Truth Table, 91
Barker Sequence, 81
Binary Sequence Filter, 4, 48, 100
Binary Store, viii, 1
Binary System, 1
Biphase Modulation, 97
Biquadratic Residue Sequence, 79

Cauer Synthesis, 7
Codes, 95, 96
Coin-tossing Sequence, Waveform, 59–62
Complementary Function, 12
Complete State Diagram, 18, 91
Complete Visit, 87
Complete Walk, 87
Congruence, ix
Continuous Autocorrelation, 61
Convolution Theorem, 103
Correlation Technique, 102, 103–107
Coset, 54–56, 110, 111
Cross correlation, ix, 64
Cross Talk, 100
Cycle, 16, etc.
Cycling Register, 22
Cyclotomic Polynomial, 41

Decimation, 52
Delay Lock Tracker, 97
Delay Operator, viii
Delta Modulation, 100
De Morgan's Theorem, ix
Detached Exponent Notation, viii
Difference Set, 75–78
Dirac Delta Function, x
Discrete Autocorrelation, 61
Division, Mechanisation of Polynomial, 47
Doubling, 87
D-type Flip Flop, viii

Euler Function, ix, 22, 23
Exponent, Polynomial, 28

Factorable Polynomial, 32
Fermat's Theorem, ix, 25
Fermat-Euler Theorem, ix, 38
Fourier Analysis of *m*-sequence, 66
Frequency Division Multiplex, 101
Frequency Domain Analysis, 102
Full Length Sequence, 16, 51, 72, 85, 86

Gray Code, 88

Hadamard Design, 75–78
Hall Sequence, 80
Hamming Distance, 80

Incidental Modulation, 97
Inverse Repeat Sequence, 82

Jump Technique, 50

Karnaugh Map Technique, 84
Key Sequence Theorem, 88, 89
Key Word Theorem, 88

Linear Feed Back Shift Register, 27
Linear Logic, 1, 3

Majority Logic, 100
Maximal Polynomial, 28
Maximum Length Sequence, 16, 28, 62
Memory Element, viii, 1
Minterm, 93
Mobius Function, ix, 22
Modulo-2 Addition, viii
Moment of Sequence, 110
Multiplexing, 100
Multiplier, 52, 53, 76, 110
Multivariable Systems, 109
Mutually Orthogonal Sequence, 83

NAND function, vii
Noise-like Property, 58

Non-linear Feed Back, 89, 91
Non-maximal Prime Polynomial, 30
NOR Function, viii
NOT Function, vii

Octic Residue Sequence, 79
Optimum Code, 95
OR Function, vii

Particular Integral, 12
Perfect Sequence, 53, 63, 72
Periodicity, 16
p-nary Sequence, 80, 109
Power Spectral Density, x, 61
Preference Rule, 85
Prescribed Period, Sequences of, 48
Prime Polynomial, 55
Product Sequence, 46
Pseudo-random Sequence, 58, 64

Quadratic Residue Sequence, 54, 73
Quasi-Linear Logic, 3, 18
Quotient, Significance of, 30, 48

Radar Range Measurement, 96–99
Realisability Restriction, 9
Remote Term Calculation, 44, 45

Sampling, sequence, 52–57
Separate Cycles, Condition for, 20, 90
Sequence Truncation, 50
Shift and Add Property, 15, 43
Shift Register, 1, 15, 16
Sonar, 97
Spectral Distribution, 70
State Transition, 18
System Drift, 108, 110
System Noise, 107
System Parameters, evaluation of, 102
System Transfer Function, 102

Time Domain Analysis, 102
Time Moment of Sequence, 109
Truth Table, 1, 84, 89
Twin-Prime Sequence, 80
Two-level Autocorrelation, 53, 59, 63
Two-level Sequence, 72, 78

Uniform Binary Code, 95

White Noise, 60, 61, 106
Word Detector, 50